Praise for Richard Manning and *Grassland*

"In elegant and elegiac language, Richard Manning has limned a map which will guide us back to our roots and help us to see the stark, insistent beauty of the grassland and understand its imperatives as never before. *Grassland* is a stunning book about the prairie, its botany, its creature population, and its remnant human population that calls it heaven." —Mary Clearman Blew

"A great work of American history. *Grassland* should be required reading for any schoolchild, and CEO, and any and all consumers. This book reminds me of the work of Wallace Stegner. It is great, necessary, and vital thinking, and I thank Dick Manning for it."
—Rick Bass

"A triumph. America's grasslands may finally have found their John Muir." —Frank J. Popper, Rutgers University

"*Grassland* is a book about almost half of the United States. It tells a vivid if sometimes deeply troubling story of loss and misuse. It is an important book." —William Kittredge

"A superb job . . . well-written and thorough. . . . This is an expansive book, one that admirably matches the vastness and sweep of the landscape discussed." —*Wild Earth*

"Here is a book where the page leaves extol the virtues of the leaves of grass most splendidly." —Wes Jackson, President, The Land Institute

"A heartfelt evocation of the prairie ecosystem that we have nearly lost and a hopeful call for its resurrection."
—*The Missoulian*

"Grasslands are the original Eden of human beings—our home. In *Grassland*, Richard Manning explains America's lost paradise and inspires us to d[...]o restore the prairie ecosystems that still re[...] and vital book."
[...]ick Smith

[...]*sland* prods the mind and stretches the
[...]se Erdrich

ABOUT THE AUTHOR

Richard Manning is the author of *Grassland*, *A Good House*, and *Last Stand*, a finalist for the Sigurd F. Olson Nature Writing Award. He worked as a reporter for fifteen years, including four years at the *Missoulian*. A recipient of a John S. Knight Fellowship at Stanford University and a three-time winner of the *Seattle Times* C. B. Blethen Award for Investigative Journalism, he has also won the Audubon Society Journalism Award and the first Richard J. Margolis Award for environmental reporting. His work has appeared in a variety of magazines and newspapers, including *Harper's*, *Audubon*, *Outside*, *Sierra*, *E*, *High Country News*, and *The Bloomsbury Review*. Richard Manning lives in the house he built with his wife in Lolo, Montana.

Grassland

The History, Biology, Politics, and Promise of the American Prairie

RICHARD MANNING

PENGUIN BOOKS

PENGUIN BOOKS
Published by the Penguin Group
Penguin Group (USA) Inc., 375 Hudson Street, New York, New York 10014, U.S.A.
Penguin Books Ltd, 80 Strand, London WC2R 0RL, England
Penguin Books Australia Ltd, 250 Camberwell Road, Camberwell, Victoria 3124, Australia
Penguin Books Canada Ltd, 10 Alcorn Avenue, Toronto, Ontario, Canada M4V 3B2
Penguin Books India (P) Ltd, 11 Community Centre, Panchsheel Park, New Delhi – 110 017, India
Penguin Books (N.Z.) Ltd, Cnr Rosedale and Airborne Roads, Albany, Auckland, New Zealand
Penguin Books (South Africa) (Pty) Ltd, 24 Sturdee Avenue,
Rosebank, Johannesburg 2196, South Africa

Penguin Books Ltd, Registered Offices: 80 Strand, London WC2R 0RL, England

First published in the United States of America by
Viking Penguin, a division of Penguin Books USA Inc. 1995
Published in Penguin Books 1997

12 13 14 15 16 17 18

Grateful acknowledgment is made for permission
to reproduce the map on page xi,
which is from *Grasslands* by Lauren Brown.
By permission of Alfred A. Knopf, Inc., and Chanticleer Press.

THE LIBRARY OF CONGRESS HAS CATALOGUED THE HARDCOVER AS FOLLOWS:
Grassland: The history, biology, politics, and promise of the American prairie/
Richard Manning.
p. cm.
Includes bibliographical references.
ISBN 0-670-85342-9 (hc.)
ISBN 0 14 02.3388 1 (pbk.)
1. Grassland ecology—United States. 2. Grasslands—United States.
3. Grassland conservation—United States.
I. Title.
QH104.M37 1995
574.5′2643′0973—dc20 95–10073

Printed in the United States of America
Set in Adobe Sabon
Designed by Ann Gold

For bison,
for all that created them
and for all they create

Gird up now thy loins like a man; for I will demand of thee, and answer thou me.
Where wast thou when I laid the foundations of the earth? declare, if thou hast understanding.

—Job 38:3–4

The world has no name, he said. The names of the cerros and the sierras and the deserts exist only on maps. We name them that we do not lose our way. Yet it was because the way was lost to us already that we have made those names. The world cannot be lost. We are the ones. And it is because these names and these coordinates are our own naming that they cannot save us. That they cannot find for us the way again.

—Cormac McCarthy

Contents

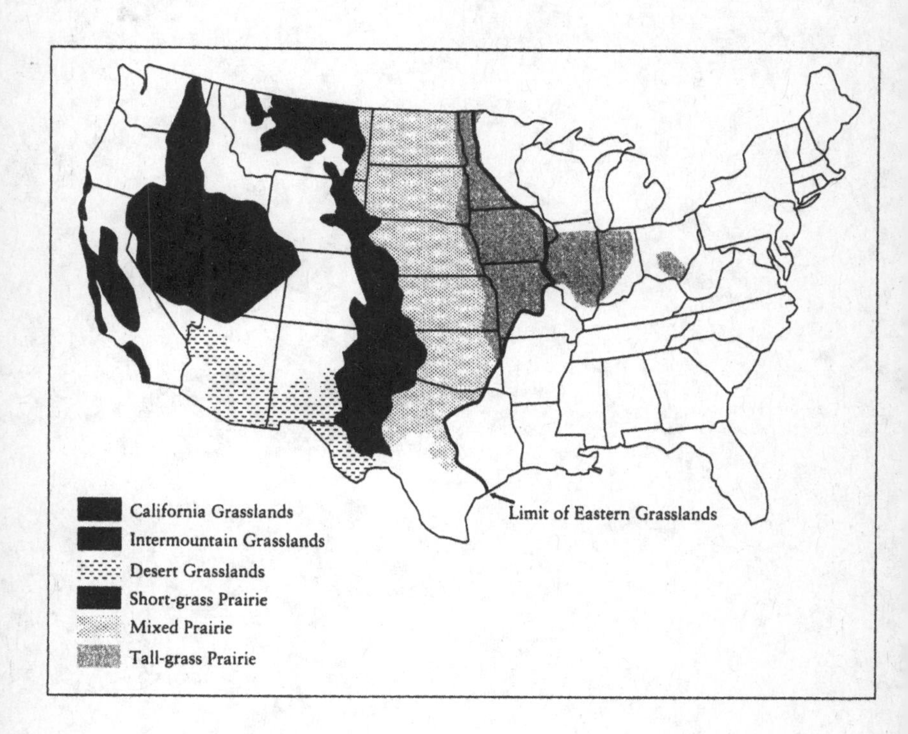
California Grasslands
Intermountain Grasslands
Desert Grasslands
Short-grass Prairie
Mixed Prairie
Tall-grass Prairie
Limit of Eastern Grasslands

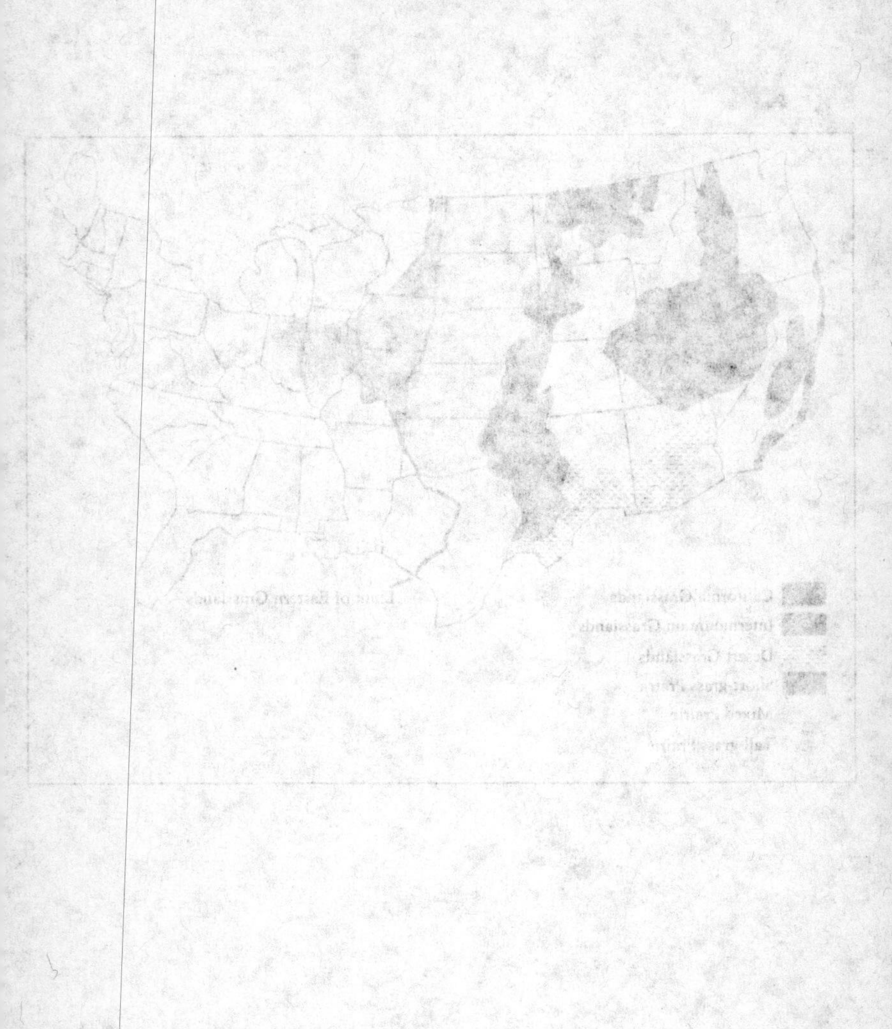

GRASSLAND

1

The Promise of Grass

This story of grass ranges like the prairie's hills, and like the hills, is best taken from the small to the large, from the specific to the general, and from the material to the spirit. In the end, though, it is a story of spirit.

There is no reason to write a book unless the process of imagining it changes one's life forever. This one changed mine, or maybe just explained some changes already made. Who knows for sure? After a life mostly spent among trees, I now understand that my life seems comprehensible only when I place it in the context of grass. I am not alone in this. We as a nation and as individuals are much defined and explained by the great empty middle of our continent, which is the grassland.

Grass has relevance to each of our lives, although almost none of us lives in the sweep of plain between the Mississippi and the Rocky Mountains and the smaller grassland domains west of the Rockies. We are all creatures of grass, if only because grassland defines a world that we are not and so defines us like the black defines the day. It is silent; we are not. It is free, and we aren't. It is large to a degree we cannot comprehend, so much so that we as a nation have spent 150 years in an assault on its whole, trying to reduce it to bits that fit our grasp. Still, stripping off all its grass does not make it less a grassland, though it makes us less of a people. Grassland is indivisible. It endures. We, as now constituted, cannot.

The notion of grassland describes what ecologists call a biome, which is a broadly defined type of habitat sharing certain characteristics. The earth's terrestrial biomes are tundra, forest, desert, and grassland. There are no firm borders separating these. Biomes weave in and out of one another in a variety of combinations and permutations. Sometimes one contains in miniature examples of all the others. Still, we can identify clear areas on the globe that are grasslands: the sub-Saharan plain of Africa, the pampas of South America, the steppes of Eurasia, much of Australia, and the American Great Plains.

Each of these is arid, receiving between ten and thirty inches of rainfall a year. Anything drier is a desert; anything wetter, a forest. This aridity is first the defining and implacable factor of grass. Plants survive by as many strategies as there are plants, but generally adapt to drier conditions by becoming shorter. This is mostly why arid lands will not support trees in sufficient numbers to become a forest, that and the corollary condition that dry land burns frequently and fire kills trees but not grass.

The grasslands of the American West are created by the Rockies and by the coastal mountain ranges of California and the Pacific Northwest. The mountains impose rain shadows downstream of prevailing westerlies, because the rise of warm air along the windward side of the mountains wrings moisture out of passing weather systems. This makes the aridity that makes the treeless plains.

We have a variety of words for this place, beginning with the French word "prairie," which originally applied to a sort of lawn. "Lawn" is too puny a word to describe the grassland, but it was the only tool the French had to work with at the time. Strictly speaking, the word "prairie" now applies only to the tall-grass, or true, prairie, the easternmost tier of the total American grassland. The true prairie runs in a band with an eastern edge roughly bounded by the Mississippi River, but its fingers cross the river and reach east, penetrating into the trees as far as Ohio. The tall grass was just what the name said, species that grew taller than a horse, such as big bluestem and Indian grass. Virtually all of it is gone, plowed under by European settlers who now raise corn, which is nothing more than a domesticated tall grass. The remaining examples of prairie exist mostly in yard-size swatches maintained as curiosities—botanical zoos—much as the California condor exists in wire cages.

The tall-grass region is flanked on the west by a stripe of mixed-grass prairie, running from Saskatchewan into Texas. Because the effects of the rain shadow become more severe closer to mountains, the grasses become progressively shorter as one travels west. There is no fixed boundary between types; sometimes the edge shifts east or west hundreds of miles, varying with drought cycles. Some species of tall prairie grasses extend into mixed prairie. Likewise, some mixed or mid-grass species extend west to the third band of grass that borders the Rockies, the short grass. More of the original short grass and mixed grass survives than does tall grass, although much of this has been converted to the raising of wheat, itself a domesticated short grass.

Some biologists subdivide the short-grass region into two groups: the high plains and the short-grass steppe. The latter is an island of specific species in Colorado and Wyoming, with the high plain wrapped around. Both areas are dominated by short grasses.

The nation also holds large grasslands in the western rain shadow that is the Great Basin between the Rockies and Sierra, farther west in California, as well as smaller bands in Florida and isolated plots in the East. Altogether, the grasslands cover about 40 percent of the United States, making it the nation's largest biome, but also our most degraded and most misunderstood.

One learns to read the world of grass by tracing its various species. Grasslands have cultures, or races of inhabitants that vary with the surrounding land. The cultures begin and manifest themselves in the species of plants, but reverberate throughout the animal world, including the human world. Grass is a map of the landscape that reveals the land's secrets, its subsurface, its seasons, its time, and its inhabitants.

In this sense the Rockies, for instance, are best seen not as a wall but as islands in a sea of grass. Just west of the stretch of the Rockies where I live, there is a place called the Palouse Prairie, once a grassland, now a wheat land of eastern Washington and northern Idaho. The bunchgrass species that defined that prairie are distinct from those of the better-known grasslands of the Great Plains, east of the Rockies. Yet the Palouse is not hemmed by the Rockies. The species of grass characteristic of the Palouse finger through the intermontane valleys of the western Rockies and spill out across the Great Divide. Some Palouse species run well east of the Rockies, onto the Great Plains, where they finally intermingle with another culture of grasses, distinct to the high plains.

This mixing of species is a record of the climate shifts during glaciation, evidence that the places that now appear geographically discrete once were connected. Eventually one learns to read the evidence of the communities of grass like a history. Today the mix of species in our country includes members from Africa, Asia, and Europe, the result of human travel and our penchant for meddling. The grass contains a record of the way we live our lives, the sort of rambling record that makes this story run far away from the place where it began, and organizes the telling of this book. There is more at stake here than a concern for the integrity of the landscape; this story has its roots in a concern for the integrity of one's culture and one's own life.

One can approach grasslands as if considering a place, but one soon realizes the definition includes a condition—aridity—along with process, or processes, mobility and its acceptance, insecurity and uncertainty and their acceptance, and ultimately a state of mind called freedom. All this makes one realize that the telling must proceed as the West does, in broad and jarring sweeps of space and time. Sometimes the story flows in details as fine as a single blade of grass, sometimes with the sweep of a whole plain as it is drawn from mountain range to range, from Ice Age to just last week. The story covers a lot of ground and so must be told through journeys, the journeys that form a sort of mental nomadism we call journalism.

The journey of this book follows my journey. That is the duty of a writer, to go and to report. I begin telling where I began traveling, with a simple idea that formed questions that in turn formed journeys. The idea is this: Our culture's disrespect for its grasslands has produced an environmental catastrophe. It will be the best measure of the maturing of the American environmental movement when it begins to understand and combat this destruction. Further, in the process of the reversal of this catastrophe lies perhaps our strongest hint as to where we must go in the larger sense, not as environmentalists, but as humans. Our goal must change from preserving nature as separate from humans to the more necessary task of remaking ourselves so that we might function as a part of nature. Some humans are at work at this fundamental challenge just now, and I will tell some of their stories.

This issue of separation from nature traces the course of environmentalism and was, ironically, what steered it away from grasslands. The conservation movement began with accepting the contention that man is separate from nature. In the latter part of the nineteenth century, America hatched what was to become one of its best ideas by setting aside areas from development, our first national parks. Fittingly, this drive toward preservation came from industrialists reacting to the excesses of industrialism. Those first preserves were about as far removed from industrial activity as could be arranged. Our first national parks were Yellowstone in Wyoming's corner of the Rockies and Yosemite in California's Sierra. They were preserved as scenery, and the grasslands were not considered scenery by the descendants of the European tree culture that ruled the nineteenth century. Parks were, first, places to rest one's eyes from the ugliness of our cities. They were simulacrums of nature roped off like a rich man sequesters his art collection.

Over the years, we developed a concern for the life of these places as distinct from the scenery of these places, and the life was mountain life. First the fauna and then the flora came under our protection, and so each park evolved as a sort of scenic zoo. In early-day Yosemite, elk and deer were kept and displayed in pens. In recent-day Yellowstone, bleachers ringed night-lighted garbage dumps for the observation of bears.

Our understanding of wild places, however, has matured. Slowly, by watching these places, by coming to love these places, we began to understand that the web of life here is extraordinarily fragile and damaged by the least of our meddling, so the environmental movement, at least in the West, became the wilderness movement. Slowly, it moved and still moves away from setting aside wilderness solely for scenic values (although that still is done) and toward setting aside places that are free from human meddling. In these vast places, we might best understand nature. We could observe and preserve elk, mountain goats, and grizzly bears and all of the twists of life that support them. The movement spread east, but well east, skipping the grasslands, to the forests of the coastal ranges east of the Mississippi, again to mountains and forests, and so at bottom, the environmental movement became a mountain-and-forest movement.

In my home in western Montana, the movement is specifically a battle with loggers. People go to jail, bury themselves to their necks on logging roads, spike trees, and hire lawyers, all in the worthy cause of preventing just one more ridge from being stripped of its trees. This is our service to wild life. This is how we have come to hate a clear-cut forest, because a clear-cut wipes out the intricate web of forest that supports all of the life in it. From this understanding, the movement has created or appropriated terms like "biodiversity," "polyculture," "biological services," and "sustainability" to explain the intricacies of the forest.

Now, though, we are coming to know that this web of life holds humans. Preservation is no longer an act of separating nature from humans. This understanding takes us to the grass.

No one I know, not even the most calloused of loggers, would consider a clear-cut pretty. No one would, for instance, hang a photo of a logged hillside on his walls to brighten his days. Yet people hang photos of wheat fields and speak fondly of amber waves of grain. A wheat field is nothing more than a clear-cut of the grass forest. Just as a forest is not only trees, a grassland is not only grass. It is hundreds, literally hundreds, of species of plants woven together in a complex fabric of interdependencies that extend then to insects, to birds, to a carpet of rodents, to predators, and finally to large mammals, of which humans are but one. A grassland is a short forest, or more likely was a short forest, in that most of the grasslands of our nation have been lost to row cropping or seriously compromised by generations of overgrazing. Even the most well-meaning of grassland inhabitants believe they do good by planting imported, exotic trees to replace native grass, a practice every bit as destructive as planting a wheat field.

Meanwhile, the study of forests and mountain wilderness has produced some curious realizations. The big fights now in wilderness, back in the mountains, are for preservation of what some biologists wryly label the "charismatic megafauna"—big, glamorous mammals that look splendid on the fund-raising posters of environmental groups. These are the creatures that capture our imaginations and at the same time are the most seriously threatened by human encroachment: the grizzly bear, the gray and Mexican wolf, the bison, the elk. In many cases, we fight for wilderness in the mountains solely to establish their preserves. Yet each of these creatures lives in mountainous forests only as an adaptation to encroachment on its original grassland home. By preference, grizzlies, wolves, bison, and elk are creatures of the plains, creatures of grass.

The difficulties extend beyond megafauna. In recent years ornithologists have become vexed by the decline of certain migratory songbirds. Because many of these birds winter in tropical rain forests, it was at first thought that deforestation in Central America was sponsoring their demise. Eventually we learned that agricultural practices on the North American plains, where the birds nest in summer, were doing as much, even more damage. Many of these birds were creatures of the grass.

Damage has been done, and this is the beginning of the story, yet if we are to take our understanding of nature to its full promise, we cannot stop the story here. A mature environmentalism rejects industrialism's imposed dichotomy of man against nature, the idea that we damage nature to serve ourselves. That has it backward. If there is any truth to the assertion of our connection to nature, it is that when we damage nature, we damage ourselves. And when we misunderstand nature, we misunderstand ourselves.

One of our greatest periods, as a people, of groping for our identity was the latter third of the nineteenth century. By then the Civil War had wiped the naive smirks from our optimistic faces. It was a war for industrialism, and the North's victory established that the coarse, burgeoning capitalism of the time would overcome the agrarian feudalism of the South. To support this industrialism, we pursued Manifest Destiny and the industrialization of the American grasslands. The ideas that sprang to the fore following the Civil War were directly responsible for exterminating nature in the biome that covered 40 percent of the country's surface.

Walt Whitman was a nurse during the Civil War, so he sharpened a poet's senses on the din of halls full of amputees. He was a sort of midwife at the birth of the new child of the nation. Fittingly, when he wished to write about democracy, he wrote *Leaves of Grass.* Grass was a metaphor for the sort of radical democracy he proposed. Yet Whitman, a man of the East, an inhabitant of New York City, knew nothing of grass. When colonists on the East Coast leveled the forests, they imported grass seed from the Old World for the pastures to feed their cattle. When Whitman made his first trip west by train to Denver, the poet of the grass was so stunned by his first view of real grassland that he wrote of "the tree problem," proposing that the land be planted with trees and peopled with yeoman farmers as rapidly as possible.

Whitman was a romantic, and curiously, most of our tradition of nature writing stems from his school. The romantics gave us Thoreau and Muir and a tradition that continues to this day, a tradition that ignores the nitty and gritty of nature and instead skims the grand and beautiful from nature's surface to cast a better poem, a tradition that, in the words of Wendell Berry, uses nature as a "reservoir of symbols."

Grassland is not symbol; it is real. Its relationships are real, which in effect makes Whitman's assertion all the more pertinent to our time. He was right about the connection between democracy and grass. Real grass and our disregard for it have much to say about freedom.

The nation settled its grassland during the adolescence of its postwar industrialism. I write this book because I wish to revisit Whitman's assertion by returning to the American West. I write it because I believe the grassland was where we destroyed democracy because of our inability to accept and understand freedom. We enter now a world that some call post-industrial, which is a world in need of freedom.

By revisit, I do not mean "re-live." The nineteenth-century West of the modern imagination is gone and probably never existed. The journey this book begins is not a fight against the entropy of time. There is nothing to be gained by dressing in rawhide like an accountant at a mountain-man festival, in a Stetson and boots like a drugstore cowboy, or flocking with the New-Agers to ersatz Lakota sweat lodges. By revisit I mean make the mental journey to patch some holes in our understanding and the physical journey to a future informed by this new understanding.

There is, after all, the possibility of a richer future. That is the promise of grass, and I mean this literally. The grass can live again. Unlike forest, it recovers rapidly, some of it in a matter of five or ten years. There is a possibility of resurrection in a real and physical sense. There is a possibility of bison and wolves.

Real and physical, that is the main thing, because it is my intent to break here cleanly with the romantic tradition that raised my writing. The grass I write of is not metaphor, nor is the West that holds it. Nor are the journeys. Grassland is at bottom a place of journeys.

Many times in the course of preparing the following pages, I traveled out across the plains. I would follow the mountain highways that weave along the river valleys as they broaden and break to the plains, both east and west, leaving this mountainous archipelago for the sea of grasslands. Over the years, this aggregation of journeys coagulated as a single voyage that pulled my life across the line I wrote of in opening this discussion. I realized I am a man of the grass.

I set out the research of all of this wishing to write a book of science, politics, and journalism. I did not plan a process that would result in that opening page, reeking as it does of the first person, but that's the way it went.

2

Forsaking the Sweet Grass Hills

There is no explaining the peregrinations of the elk named Earl. The more closely we examine the journey, the more clearly we see the conclusion that each of its 1,800 miles only illuminates our ignorance of the parallel universe we call "animal."

We can understand only the most rudimentary of the facts about this case, but these are clear enough. One day the animal stood up and walked away from a safe home in Montana's Sweet Grass Hills, walked just across the border into Canada, maybe 80 straight miles to Alberta's Cypress Hills, then walked to the Milk River, then back south to Montana, then walked south-southeast, sometimes due east, maybe, no one knows for sure, maybe along the Missouri River, but for sure to Independence, Missouri, 1,800 miles away.

We know this because in 1987 a wildlife biologist and plainsman of some experience slipped a collar holding a radio transmitter onto the neck of this yearling bull elk. The beeper conked out, but the collar still bore a number that was the animal's positive identification three years and 1,800 miles later.

Number 964—until the animal became something of a plains-state celebrity at the end of its trip—was its only name in the notes of the biologist Gary Olson, a Montanan employed by the state to attend to wildlife (by which the state means game herds) that has survived the taming of the plains.

The collaring was a normal enough event in Montana, which holds about 75,000 elk all along the western edge of the state, spread on the mountain peaks in summer, across the mountain valleys and foothills in winter. The fact that they live almost solely in the Rocky Mountain Range would make it seem that the elk is a mountain dweller, but it isn't. Elk once ranged over virtually all of the continental United States save a very small strip in the extreme Southeast. No elk have been recorded in Florida, for instance, but in the 1770s the naturalist William Bartram observed them in Georgia. New York had them when the colonists arrived. So did the Potomac River valley.

The colonists gave the animals the name "elk," a misnomer drawn from a lineage of words that in most European languages refer to the moose. This was because the colonists from England did not know of the great stag of Continental Europe, the Eurasian red deer. This red deer is what an American elk is, biologically the same as the European form. Some people prefer the elk's Indian-derived name, "wapiti," a term popular in hunting magazines, but to biologists the names red deer, elk, and wapiti all mean *Cervus elaphus*. There are six American subspecies. Like humans, all elk first came to this continent by way of the Bering land bridge during a recent ice age, migrants derived from Old World stock.

Despite their present concentration in mountains, the clearest attribute of elk to emerge from their record of at least 120,000 years on this continent is that they are creatures of grass. They eat grass first, along with the attendant leafy plants of grasslands. Unlike deer, elk eat shrubs and trees only occasionally, when forced. The status of grass being what it is, though, they have been forced. The grasslands are gone. The Great Plains to an elk look dead as a hammer. This, not the elk, is the real issue of this story, but the elk is the first thread of its telling, the elk and the trip from the Sweet Grass Hills.

Despite their once far-flung range to places like Georgia and New York, elk appeared only occasionally east of the Mississippi, always in spots coincident with a few fingers of grass that broke the broad canopy of forest. They were anomalies in the East. Farther west, though, as these fingers traced back to that great muscular hand of grass that is the plains, the elk thrived until hunters and plows put them down.

In Indiana they were gone by 1840, in Illinois by 1850. They were abundant in Kansas until 1871, then faded. In 1976, someone found thirty elk in the state. Where they came from, no one could say. In Missouri, a state of particular relevance to our story, elk were exterminated by the turn of the century. In 1876, *Forest and Stream* magazine reported on the care of elk in Missouri:

> Owing to a savage and indiscriminating warfare that has been inaugurated against them within the past few years, their numbers are decreasing more rapidly than ever before. . . . An elk skin is worth from $2.50 to $4, and to secure that pitiful sum this beautiful life is taken and the 300 to 500 pounds of the most delicate meat is left on the ground.

Writing in 1871, the naturalist J. A. Allen described the management of the elk of Iowa's prairie:

> They were of great value to the settlers, furnishing them with an abundance of excellent food when there was a scarcity of swine and other meat-yielding domestic animals. But, as has been the case too often in the history of the noblest game animals in this continent, they were frequently most ruthlessly and improvidently destroyed. In the severer weather of winter they were often driven to seek shelter and food in the vicinity of the settlements. At such times the people, not satisfied with killing enough for their present need, mercilessly engaged in an exter-

minating butchery. Rendered bold by their extremity, the elk were easily dispatched with such implements as axes and corn-knives. For years they were so numerous that the settlers could kill them whenever they desired to, but several severe winters and indiscriminate slaughter soon greatly reduced their numbers, and now only a few linger where formerly thousands lived, and these are rapidly disappearing. Their home here being chiefly the open country, they much sooner fall prey to the "westward march of civilization" through the most merciless treatment they received at the hands of the emigrant.

It is tempting to say the elk fled all this for a friendlier place. Mostly, though, the elk did not flee the plains; they were killed there, killed both outright and indirectly, when the grass that supported them was plowed under. The remaining body of elk coincides largely with the body of the Rocky Mountains and west to the ranges of Oregon and Washington, the tail edge of the Sierras, and a few high islands around. They have taken refuge where it is too steep to plow.

Such retreats by whole species, even before humans arrived, are not unique in the history of our continent, warped, folded, and frozen as this place has been. During the frequent glaciations that have been the rule of recent history, plants and animals retreated south or to the very tops of mountains to wait out the 100,000-year winters, and during the drought and desertification that frequently followed glaciation, life retreated to rain-catching mountaintops. Paleoecologists call the places of retreat from ice and drought "refugia."

In the northern plains, there were and are three key refugia: the Black Hills of South Dakota, Canada's Cypress Hills, and the Sweet Grass Hills of Montana.

The Rocky Mountain Front, or simply the Front Range, is a term of resonance from Canada south to New Mexico. It is the name for the eastern edge of the Rockies, a sort of hinge between mountains and plains. Those of us who have come to know this long line of place regard it with a singular reverence. To be there on a sharp black morning, to sit backed by rock and ice peaks on a prairie bluff and watch the first flush of rose mute the morning stars, then spread like the glow of fire across the grass-flattened horizon, burn from rose to orange, a full half hemisphere of night awakening to blue—to be there is to forget all envy of wealth, cleverness, and power. This place bestows blessings that obviate all these.

In such a place in Montana, the Missouri River rises.

In the northern plains, they say fidelity to the mountains and the prevailing western winds shapes humans and plants. They say this is where "the people lean west and the trees lean east." Like the people, the river that rises here doesn't seem to want to leave the mountains.

From the point the Missouri River begins as streams—such as the Gallatin, Madison, and Jefferson, mountain trout streams, headwaters named by Lewis and Clark for their bosses—its job is to define the plains, but it takes its sweet time getting down to work. The Missouri plateau is a grand, flat chunk of land bounded by a ragged-edge version of the Canadian border on the north, the Rockies on the west, the Mississippi on the east, and fingered tributaries of the Missouri into Wyoming and Nebraska on the south. These are the high plains, the Great Plains.

Yet instead of beginning in mountains and flowing straight east, as one would suppose, the Missouri gathers its headwaters and heads north, straight up the crease of the Front Range, paralleling the mountains like the pin of the hinge. From its headwaters at the Idaho state line it flows north to Great Falls, stubbornly clinging to the hills for a distance of more than three hundred highway miles. Only then does it leave the easy rolling land of aspen groves, limber pine, foothills, buffalo jumps, and buttes to string together the plains.

The Missouri's first city of any consequence is Great Falls, a dusty farm town of wide streets and low aluminum-fronted buildings, plastic signs, cafés full of sugar donuts and acid coffee. People drive pickup trucks, and the wind sharpens its bite on grit sucked from surrounding plowed fields.

The highway east out of Great Falls follows the Missouri up a steep hill and then as suddenly as an ax dumps onto the plains. Hammered flat and vast by the burden of the glaciers, the land now shoulders nothing so much as hard, red wheat. Unblemished by fence row or pasture or unworked strips of nobody's land, this place is to wheat what the Sahara is to sand. Here the Missouri ducks out of sight, down into a broad canyon sliced into the plains. From the highway, the brim of the canyon appears now and again only as a void in the wheat. Occasionally, the highway works itself to enough elevation for a teasing view down into the grassy green crotch of the canyon, a hint the river is there, but mostly the view is of wheat.

Even here, the Missouri refuses to run straight east. Out of Great Falls, its vector is still half northerly, as though it wants to avoid the plains. It runs to the town of Fort Benton as if some irresistible tug pulls it to the north, deflecting it from the flat land, as if it's looking for a place to run uphill, not down. Only just east of Fort Benton does it finally surrender to the pull of the Mississippi basin and arc away from the north. The tug that pulled it north finally lets go. On a good day, you can see this tug, and if you let it pull you off the Fort Benton highway, it will take you straight north along a paved county road with no shoulders.

Farmhouses, mostly modern and modest, dot the wheat fields about every five miles. Their mile-long driveways lead to houses, clusters of steel buildings, and hard-packed gravel lots where monstrous tractors are parked. On the highway the traffic, a vehicle maybe every half hour or so, is mostly pickup trucks. One quickly grasps the rules of this road: Drive seventy-five miles an hour and wave at every oncoming driver.

The Sweet Grass Hills emerge with increasing clarity on the horizon. They're really a cluster of buttes, volcanic rock heaved into piles seventy million years ago. At their tallest, they rise to 6,400 feet, more than 3,000 feet above the surrounding plains.

The county road runs almost straight north, to the town of Chester, which is on what Montanans call the Hi-Line, an east-west rack of northern-tier counties that stretches from the Rocky Mountains five hundred miles to North Dakota. The line that defines it is either U.S. Highway 2 or the railroad, once Jim Hill's Great Northern, then the Northern Pacific, now the Burlington Northern, take your pick. It's a lonesome place.

Maybe every fifty miles, the Hi-Line is spotted with towns like Chester, strung between the highway and tracks. Always there is a grain elevator; sometimes a school, a couple of churches, a Circle K, a Tastee-Freez, and a dealership that sells big tractors. Always there is grit and wind. Chester offers almost nothing to distinguish itself from these standards except a clear view of the Sweet Grass Hills. Most prairie towns are not blessed with hills. They seem to stand just on the edge of town, but they do not. To get there, one drives, still at seventy-five mph, straight north on a gravel road for maybe twenty-five miles. About halfway up the road lies the six-thousand-acre wheat farm of Arlo Skari, as good a place as any to begin asking questions of the hills.

The Skari ranch compound is three houses—Arlo's, his son's, and his brother's—and a clutch of steel outbuildings. Along a small creek stretches a row of trees of a wide variety of species, anything Arlo and his wife, Darlene, could get to grow here in a place nature intended as treeless. These fringes of trees around farm buildings are called shelter belts and are the plains-land equivalent of the bonsai garden, vegetation so compulsively attended as to gather religious significance.

Hunkering in the lee of the shelter belt is Arlo and Darlene's house, the newest of the three, a cedar-sided ranch fronted by a broad patio and a big organic garden. Inside, the house is trim, white, and neat. There is a television and VCR, microwave, and computer out front in the book-lined room that is the ranch's office. There are books spread on end tables in the carpeted living room. A few bumper stickers announce allegiance to the Skaris' choice of candidates in the fall election, all Democrats, some liberal. Both Darlene and Arlo are active environmentalists, members of the local Audubon and a group that lobbies to preserve wilderness. Darlene once sat on the board of one of the state's most vocal environmental groups. The garage holds a Jeep Cherokee, an Audi 5000, and a four-wheel-drive Ford pickup, all close to new. Arlo's son drives a Saab Turbo with a vanity plate that says "Tilth."

Arlo has white hair and a long, creased farmer's face. His grandfather emigrated from Norway. Arlo won't so much as comment on the weather without giving the impression he has thought about the matter a great deal. At dinner, which includes lefsa—a sort of Norwegian potato tortilla, frozen vegetables, and hamburger casserole, there is only the sparsest conversation. The radio substitutes, delivering "All Things Considered." Arlo keeps his head bowed all the time, in the manner of the sullen or the pious. He says he's not all that religious.

"We're Methodists, but mostly for the fellowship," he says. "Really we're probably Unitarian, but Methodist is as close as we could come in Chester."

He offers a tour of his farm, a jostling bump on rutted tractor roads that extend over 6,000 acres owned and paid for, another 2,600 acres leased. After the first mile-square section, there is not much different to see on the other nine sections. Save for the frizz of trees around the ranch house and the troughs of grass and brush in the two creek beds that flow through the place, the Skari empire is mostly wheat. The fields play out in relatively narrow but mile-long bands, alternating between strips of fall stubble from last summer's wheat and strips of broken, naked dirt. This displays a system of farming called fallow strips. Alternate strips—maybe a few hundred feet wide and a mile long—are planted to wheat in alternate years. The rotation markedly boosts production because of the fallow land. It also reduces erosion because it lays out the land in strips crosswise of the prevailing weather, to break the wind's purchase on the soil.

The plowed land is not as naked as it seems; on the surface lies a thin coating of last year's stubble, a mulch Arlo calls "trash." This too protects the soil from erosion. In earlier days, the stubble would have been turned under to rot, but new equipment and methods allow it to stay up top. Federal officials periodically inspect Arlo's land to see that he adheres to these methods. Failure to do so would cost him his federal price support payments.

"I'm not sure anyone realizes how much time we spend making sure that this soil doesn't blow away," says Arlo. He sits on the board of the local soil conservation district.

When Arlo first saw this land in 1947—then he was just out of military service and trained as a pharmacist—this land was more than wheat. It was the old Prescott ranch, then grassland, a broad expanse of native vegetation, albeit slightly abused by livestock. The plains around the Sweet Grass Hills were not homesteaded until after the turn of the century. Until the people that Montanans called "honyockers" brought in the plows, this land was open range. The homesteaders began a trend that continues. There is prestige in ranching, in raising cattle on grasslands, but there's more money in wheat and so each year more grassland falls to the plow.

A botanist and lifelong student of the region says that if one can find an ungrazed stretch of native high plains, one can identify as many as 250 species of plants inhabiting a single site. On a site that had been grazed that count would drop to about 40 species. With work, a plow, and chemicals, a wheat farmer drops the count to one species. On grazed land, soil erosion is virtually nonexistent. On wheat land it is constant.

"I have mixed emotions on whether this land ever should have been broken out," says Skari. A few miles away, a range of restored native grassland supports two hundred mule deer per square mile in winter, at times as many elk, but that place seems a long way off from this stretch of land, covered with a great tarpaulin of wheat.

The quote of a certain scripture will quickly pull a serene sort of grin across the Skari face, as if he understands viscerally what the

Psalmist intended when he lifted his eyes to the hills. The hills are a refuge from the plains. When Skari speaks of an acquaintance in a town a hundred miles across the plains he points, as if the listener could see the town and make out on the horizon the face of this friend. To look out across the plains is to swear this is so. In a place so vast, the mind can survive only by compressing it to the intimate.

How intimate, then, are the Sweet Grass Hills that seem to perch on Skari's north fence line like gods? They are not that close, but they seem it. How they must organize a life.

"When I first saw the hills, I thought they were the most beautiful thing in the world," says Skari, and he still does. The hills gave his environmentalist leanings an almost religious conviction, as if he labors for this bit of heaven on earth.

Nowadays, though, defense of the hills means Arlo must go calling. A shy man, he does more than his share of traveling and visiting, like a salesman. He is selling an environmentalism that doesn't come easy to the cussed and craggy ranchers that ring the hills, but come it must. A mining company has proposed a fresh new gold mine atop the east set of buttes, the perch that best overlooks the plains. Gold miners in the West these days do not use mules, pick axes, sluice boxes, and pans. They use Euclid dump trucks and strip-mining shovels to literally rip off the tops of mountains.

At first, says Arlo, there was some resistance to his environmental appeal among his neighbors, but the lifelong residents of the Sweet Grass Hills now are signing up to fight the mine. "They thought the hills were going to be there forever," says Arlo.

We—Arlo and I and two box lunches packed that morning by Darlene—bounce north in the pickup along the rutted mud road, now and again crossed by the barbed wire gates of the range. It is a county road and the only one that circumnavigates the hills, a day's drive. In Montana, to travel the public right-of-way is to learn to handle gates.

We are in ranching country, grassland that begins just north of

Arlo's ranch. We have crossed the frontier from wheat to grass. The human pressure on the hills and on the elk of the hills lies in sort of concentric circles around them. The biggest, outer circle holds wheat, but as the land begins to undulate and rise to meet the hills, it forms the second ring, where cattle graze, and this land is protected from wheat by virtue of its being too steep to plow. We cross land grazed hard, "grazed flat as a kitchen floor," Arlo says. He reads this land as a sign of a particular rancher's troubles: perhaps his land isn't yet paid for, or the rancher overextended himself on a new hay baler so he bled a few extra dollars from the land by overstocking the range.

We cross a fence line now to the land of a different owner, and the grass is taller, even though it is fall. We have entered the Meissner place. Arlo says the Meissners—six German immigrant brothers—have so much land it would be hard to imagine overstocking it. There are plots of Meissner land stretching a hundred miles west to the Rockies, one hundred miles south past Fort Benton. They own hundreds of thousands of acres.

We are driving onto the home ranch of this empire, the residence of one of the Meissner brothers. I expect that this much wealth must generate a palatial house. Instead, there is a vintage 1960s mobile home, not even a double-wide. Why spend all that money on a house when a trailer works fine? The brother is not home. It is daylight and daylight is for working.

"The Meissners have learned the secret to ranching," says one area observer. "You live poor and die rich. They didn't even allow one another to get married."

In their sixties, two of the six brothers broke the bachelors-only rule, but they say it was a tussle.

We continue north a bit and then the road arcs around the northern edge of the hills. Farther north still, the plain stretches off into Alberta. We stop to visit with a young rancher whose neighbors and a line of cattle trucks have gathered at his corrals. It's roundup time, and the corrals are bulging with cows. The business at hand is to separate each cow from her calf. The calves, spring issue and fattened on the summer grass of the Sweet Grass Hills, will be loaded into the trucks and sold to feedlots in places like Iowa, where there is corn. They'll be further fattened on grain and butchered.

The cows left behind will be bred and sent to bellow for their lost calves for a week or so until they finally settle down to huddle against the hard, high country winter before another run of the cycle. The young rancher is friendly but too busy to talk with us, so we head on.

We next visit a young rancher raising a flock of turkeys in her yard, mostly a job of keeping the owls away. She says she dearly hates to shoot the owls but reminds us from beneath the brim of her National Rifle Association baseball cap that she'll shoot if she has to. She and her husband manage this ranch for an absentee owner, a man raised here who inherited the place but apparently not much of an attachment to it. He has moved away and lives in a Rocky Mountain ski resort town, where there are people and more exotic ways to spend his money.

The ranch is for sale, whole or in pieces.

"He's slicing it up like a roast," the woman says, then sends us up the hill to visit old Mrs. McDermott.

Ruth McDermott is ninety-four and refuses to leave her crackerbox of a house, shelved against the toe of the Sweet Grass Hills' westernmost butte. She was born nearly in sight of this house and here she will die. Her son, Pete, still takes care of the ranch and a neighbor lady comes in most days.

We were steered to Mrs. McDermott because somewhere along the line I had been introduced as a writer. People around here assume that the past is the only thing worth writing about. Helpful people believe my business is simply a matter of finding the person with the oldest stories. Mrs. McDermott doesn't breathe so well, but still is able to sit up and talk and negotiate the few steps from the kitchen to the living room. She tells the old stories.

Like the time when she was just a girl and a drunken argument erupted in her front yard and somebody beat the hell out of Bill

Long, who in turn shot and wounded a Canadian Mountie. Then he went stalking Ira Brown, a principal in the beating.

"Ira didn't have no gun so he dug out an old double-barreled shotgun and he shot Bill Long in the stomach," says Mrs. McDermott. "That policeman stayed hid all night."

All this occurred at Gold Butte, a ghost town now, three or four miles away, but within sight of the McDermott place. Everything is in sight of the McDermott place; to the south, across the plains, mountain ranges nearly one hundred miles away shine through the clear plains air to reflect on the front window of the house the old woman won't leave.

She was raised at Gold Butte largely because her father, Rodney Barnes, found the first gold there in 1883. He made thirty thousand dollars working the hill, at the rate of nineteen dollars an ounce. The town held on into the forties. Arlo says he can still remember when the saloon, the Bucket of Blood, did a solid business next to the dance hall. Even today somebody works the gold now and again, but the town is gone. Mrs. McDermott married a miner, a man who got lucky one day when a badger dug up a few lumps of coal on his ranch. The McDermott coal mine warmed the surrounding prairie houses until 1947, when the natural gas lines came in and no one would pay the going rate of seven dollars a ton, a buck and a half to deliver, for coal.

Coal or no, the McDermotts stayed on and ranched. Pete says he made the mistake of learning to shear sheep and that kept him busier than he'd like to be most years, but it was all a part of scratching out a living.

Arlo gently broaches the subject of the proposed new gold mine with this mining family and now Pete goes somber. He's against it. In the old days men mined with a saddle horse and a shovel, but no one ever thought of ripping off the top of a mountain.

"I'd hate to see them tear up the country," he says.

Mrs. McDermott then talks about the country, how she remembers it was when she was a girl in the days before the stock came in. The country was different then, a sea of grass. You could pick wild strawberries along all the creek bottoms. Once a rancher lost a bunch of sheep up on Jackass Butte after they got into the blue lupine, a long-stemmed plant that holds a rack of pea-size, gentle blue blooms in late spring. They can make the hills look liquid as a lake. They are among the species that die with grazing. This is how the plant community becomes impoverished.

But Pete and his mother say the biggest change is recent; over the last few decades, the people have gone, and now it's a lonelier place. Around the hills, every few miles or so, there still stand a few clusters of boarded-up churches, stores, a dance hall, a cemetery up on the hill fenced against the cows. These places used to be towns. On Saturday nights there were dances, shivarees. Kids went to school in the towns instead of riding for miles on a bus, as they do now—what kids there are.

The pressures on the plains prescribe ever bigger ranches and farms. That's how a farmer makes it now, by inhaling the lands of his failed neighbors and buying tractors that work the land in a single pass. This country no longer needs its people. The remaining old folks back up to the hills, to this refugium, like plants and elk fleeing the economic glaciation below. This is how human communities become impoverished.

Lloyd Oswood is somewhere in his seventies and, by any reasonable assessment, crazy as a shithouse rat. He lives alone, as he always has, a Norwegian bachelor cowboy. His paid-off three-hundred-acre ranch lies at the inner edge of the grass circle around the Sweet Grass Hills, perched on the base of Gold Butte. He describes the place as "a-fly-by-night-catch-as-catch-can-and-hope-to-God-you-make-it ranch."

His house has two rooms, both floored with something barely recognizable as linoleum. There's a combination wood and propane stove for cooking. There's a red-painted Dutch kitchen for a cupboard, but it seems redundant since most of his staples are spread out on the table. There's no electricity for refrigeration, so a trapdoor in the entryway leads down to a cellar, where Lloyd keeps his beer.

He cautions a visitor about to brave the stairs on a mission of mercy for Lloyd: "I dislocated a shoulder going down there to get a beer. Those goddamned steps throwed me harder than any bronc ever did."

Lloyd himself is maybe five six and built like a fence post. Eyes the color of a hazed summer flash and beam above a straight hawk's beak of a nose, above a chest-length, winter-white beard and below-shoulder-length winter-white hair. The hair he disciplines forcefully with a neon orange plastic baseball cap. His jeans have a hole here and there and are capable of standing by themselves, not so much as a structure, but as a life form.

"I'm handing you no bullshit. I am," he announces, "the genuine article."

Lloyd, however, does not want to talk only about the past. Lloyd wants to talk about anything. His table is piled higher with magazines than with canned goods, an eclectic mix including the local paper, the state's stockman's news, *U.S. News & World Report,* and a mix of romances and western pulp novels.

"There's so damn many things to think about. I try to keep up, but there's so damn many things," he says. "I buy more fucking damn magazines than anybody in the country."

Then he launches into a dissertation on balloon mortgages, backed with an equal number of quotes from *U.S. News & World Report* and a romance novel he is reading. He grabs a country western music magazine.

"What do you think about this George Strait? He's a good-lookin' son of a gun. But you look here: It says he can sit a horse. He can throw a lariat. He's no panty waist."

Now we're off on a discussion of the price of bulls and artificial insemination.

Recently someone decided Lloyd should be blessed by television, so some neighbors got together and hung an old color model on the wall, then rigged up a generator to power the thing. Still, because Lloyd hasn't mastered the rudiments of turning it on, he depends on the occasional visitor to do so.

"It don't work. It ain't worth the powder to blow it up. The radio works and hell, that's just as good as television," he says. "That TV will go to howlin'. Ever have your TV go to howlin' like a soul in hell? Have you ever seen Canada TV? Canada doesn't give a damn what they say on TV. It's right out in the goddamn open."

Still, once Arlo and I get the television to working, Lloyd likes it well enough. An occasional female contestant on a game show will catch his eye and he will comment briefly on the turn of her cheek and so forth. But only briefly. Talk of women greatly embarrasses Lloyd, and he quickly changes the subject.

Mostly Lloyd is angry on this clear fall day, and Arlo and I learn this when we make the mistake of mentioning fall roundup. The neighbors are bringing in their cattle, and not a one has asked for Lloyd's help. Every time this occurs to him he stands straight out of his chair, the twinkle drains from his eyes, and he starts fuming and swearing.

Lloyd doesn't have a horse anymore. The rancher who leases his place won't let him keep one because horses eat grass and grass is for cows. But Lloyd says he can still sit a horse. If his neighbors would only do the decent thing and ask him to roundup, he'd be good help. But they don't.

"I don't know what the hell people want around here. They're gettin' so big-wheel you can't touch 'em with a stick. They don't need me, they say. They say they got hired hands."

All day Arlo and I have seen ranchers at roundup, but we've seen few horses. Because grass is for cows, the cowboy's tool now is a four-wheeled Japanese motorcycle called an all-terrain vehicle.

"Now it's those goddamned Kawasaki quarter horses they use. I'd be ashamed to wear boots," Lloyd says.

Lloyd tells us that one time an Indian woman came down off Gold Butte and walked up to his house asking for food. She had been fasting and ate so voraciously that he didn't believe he'd have enough canned goods on hand to satisfy her. This is not a matter for the big circle of wheat around Lloyd's place. Nor is this a matter for the tightening circle of ranches closer to the hills. This Indian woman had been in the center of the circle.

We give the Blackfeet Indians credit for naming the hills, and well they should bear a name of the tribe's choosing. Before settlement, the hills stood at the center of their world. The Blackfeet were in many ways the most isolated and obstreperous people of the plains. This was their land. They held to it by hunting bison.

The Blackfeet, however, called this cluster of buttes the "sweet pine hills." Whites mistranslated the term as sweet grass. The confusion was understandable, owing to the presence in the hills of *Hierochloae odorata,* known some places as vanilla grass, but mostly as sweet grass. It should not grow at all in the northern plains. A grass of wetter climates, it flourishes farther east and out of the rain shadow of the Rockies. But the extra elevation of the hills wrings water from passing clouds, giving the hills about twice the moisture of the surrounding plains, and so here the sweet grass grows in refugium.

Of course the Blackfeet knew this. Ribbons of smoke from a burning braid of sweet grass open a window to one's soul. Sage and sweet grass smolder together on the red-hot rocks of a sweat lodge, even the sweat lodges of these days, which are presided over by Vietnam vets, recovering alcoholics, and other Blackfeet people trying to deal with the complexities of this world by learning the rites of the old. The sweat lodge is a ritual of purification.

The lodge itself is an igloo-shaped mound of blankets and tarps laid over a frame of sapling hoops. The lodge is short and so in it one must sit in the dark with a circle of maybe five other people. In the center, a pit has been filled with rocks heated to their cracking point in a fire just outside. The door is closed and the dark is absolute. The heat rises and then come songs, chants, and tears, then purification through a heat so intense it cannot be held by lungs, lungs that always will remember this heat as borne by the bite of sweet grass and sage.

This sweating and crying, though, is only a beginning. It is purification, and expiation of one's sins. It prepares one for a vision.

The Blackfeet and most other natives hold that human beings are a deprived lot, that a wretchedness of spirit is the price of our cleverness. We have forgotten the order of creation: First there was spirit, then there was sun. Sun was and is the creator that first brought forth the animals. Humans came later, brought forth by a beaver that pulled us from lumps of clay at the bottom of a pond. Because humans are the farthest removed from the spirit, we rely upon animals for guidance. The ritual of the vision quest is a seeking of that guidance, a time when one goes atop a mountain to fast and wait for a vision relayed by an animal. For vision, one goes only to those mountains with a sweeping view of the rising sun and the surrounding plains.

If one knows what to look for, one can find such sites today, historic sites. Although there is a whole line of potential vision quest sites along the Rocky Mountain Front, the Blackfeet and surrounding tribes used primarily Chief Mountain (which stands in what is now Glacier National Park), the Cypress Hills, and the Sweet Grass Hills. Apparently this ritual has been going on for some time. Recently, someone found some five-hundred-year-old masks in the Sweet Grass Hills. They were made from seashells from the southern Atlantic coast.

Vision quest sites vary in construction but are generally not much more than a circle of rocks big enough to sit in. If it is an old and undisturbed site, there may be a buffalo skull nearby. A supplicant often would attach one end of a leather thong to a skull and the other end to skewers run through knife slits in the back or chest muscles and drag the skull up the mountain to the vision quest site.

Now that the bison are gone, modern sites often contain cattle skulls. Plains Indians believe in a redeemer. They believe that evidence of redemption will be a return of the bison and elk to the plains. They believe the redeemer will descend from a cloud onto the Sweet Grass Hills.

On Mount Royal, a prominent peak in the hills, there are fourteen structures: translators, transmitters, towers, and satellite dishes for broadcasting radio and television. Arlo Skari says that in a way these places in the hills are a good thing because one of them brings in National Public Radio. He says it wouldn't be possible to inhabit the northern plains without it. But he likes the wry joke that others have made about the towers being the site of the white man's vision quest.

In 1888, the Blackfeet "ceded" the hills to the whites in a treaty. In 1968, the tribe accepted a final payment of $8.7 million for their hills. They are gone, and this inner circle is now owned partly by the federal government, partly by ranchers.

The only place an elk can live today is in the inner circle, up the hills and past the wheat fields and the heaviest of the grazing land. The center is sanctuary, and it is difficult to imagine why an elk would leave it, as the one that came to be called Earl did in 1987. It is Gary Olson's job to monitor the herd of two hundred or so elk that live on the hills. He has collared, counted, and tracked them for years. He's read everything he can get his hands on, and still the journey of Earl is a mystery.

Bull elk from time to time do leave, for fear of hunters. The elk know enough to slip across the border into Canada during hunting season in the United States, but they seldom go far, certainly not 1,800 miles.

Exogeny seems a better explanation, assuming an explanation is possible or even appropriate. Many species but especially isolated populations of animals have tricks for ensuring genetic diversity. Generally, these devices involve either males or females leaving home, splitting off from the community to take up with unrelated animals. Wolves are known for possessing this wisdom, with young "dispersers" traveling as much as two hundred miles to new territory. Exogeny is thought to be the biological imperative that gave rise to incest taboos among humans.

But there is no record of exogenous behavior in elk to suggest a journey as long as Earl's. Not that the urge to reproduce (of which exogeny is a part) isn't a powerful force in the species. When inflamed by the sexual urges of the fall rut, bull elk are known to masturbate on bushes. They squeal and, with arched backs and deft aim, urinate on their own faces. The biggest of bulls, the most successful breeders, generally copulate away all of their energy during the rut, then die during the tough months of winter. Such a sex drive certainly could make one walk 1,800 miles.

Earl, however, traveled right on past other populations of elk, to a place where there were none, or where at least there had not been any for more than one hundred years. He did eventually find other elk in Missouri, but only after being drugged, trapped, and locked up in a zoo.

Olson figures the journey had something to do with grass. Food plain and simple, or maybe not so simple. It is not a great leap to believe that the grass is only a hint, a trigger of the elk's deep racial memory that was the real force behind his moving. All life is inextricably tied to the land that made it. Close as animals are to the creator spirit of the land, they would know this better than we do. Earl's journey was in a sense a following of the directions encoded in his genes to reclaim the land that made him. This elk acted as if he were reclaiming the plains.

After his brief detour north into Canada, this elk hooked back south into Montana to a path not altogether new. In a recent and curious development on the northern plains, elk are showing up frequently well east of the Rockies. This is where grass comes in. In the early eighties, wheat farmers who had started going broke arranged for a federal subsidy by arguing that the money would halt soil erosion. This argument turned out to be a good one. During the farming boom of the 1970s, plows had made their way onto increasingly marginal and erodable sites. A decade later, soil conservation officials had become so alarmed by the increase in erosion, they were willing to go along with some federally subsidized retrenchment. Under the 1985 farm bill, farmers could idle a portion of their wheat land by planting it to grass. In turn, the federal government would pay an average of thirty-five dollars per acre per year for land so idled. By the end of the decade, farmers in Montana had enrolled about 2.5 million acres in the program.

Critics argue the program could have gone much further toward its stated goal of conservation. For instance, it contained one massive loophole abused by many farmers. That is, farmers with virgin prairie simply idled wheat land by planting it to grass, took the subsidy, and plowed up the native grassland to plant more wheat. Some farmers planted their idled acreage to crested wheat grass, an exotic and easily grown species of grass that is useless to wildlife. The program had the net effect of creating a patchwork of grass across the prairie, often monocultures of substandard grass, nothing that resembled the carpet of plant life that once blessed the place, but at least there were grassy oases in the desert of wheat. Olson says the effect on wildlife has been enormous. Even this little bit of restoration has set the elk wandering again out across the plains, as if these poor grasslands suggest to them what might be. Elk now appear routinely a couple of hundred miles east of the Sweet Grass Hills.

None of them, however, have done what Earl did, and there is nothing in these few patches of grass to explain fully why he did it. Maybe the new grass raised possibilities in the elk brain. Almost nothing is known about his trip after he left the Cypress Hills. Olson guesses, but allows it is only a guess, that the animal soon found his way back to the Missouri River and simply followed it downstream. The Missouri would provide a break in the unrelenting tyranny of the wheat. Its course mostly is lined with cottonwoods, brush, and grass, places where a bull elk could feed and hide.

Its course, however, also is dammed by reservoirs and crossed by bridges, railroad tracks, irrigation diversions, subdivisions, and interstate highways, more and more of them the farther one goes downstream. Still, the elk started near the Missouri River, and that's where he ended, so it's hard to imagine him straying too far from what sometimes seems the only natural feature in an artificial landscape.

There was absolutely no news concerning the whereabouts of the elk for two years. Then in 1989 there came a few scattered reports of what could have been this elk, from some rural counties north of the Missouri River near the city of Independence, Missouri. In 1990, the animal went urban and began showing up with considerable regularity south of the river. Frequently, he was misidentified, urban culture's distance from animals being what it is.

A report in March of 1990 in the *Independence Examiner* said this:

> Have you seen an unidentified roaming antelope in your neighborhood lately? The Missouri Department of Conservation has received numerous reported sightings of an antelope wearing a white collar around the 192 highway and the Independence Center [shopping mall]. Other reports have speculated the animal was a moose or elk.

Later, the *Kansas City Star* picked up the story:

> Independence's mysterious elk was trotting north down Adams Street toward a garage sale about 8 A.M. Wednesday when an impatient driver honked his horn.
>
> The startled animal took off in such a hurry that he left foot-long divots in Woodford and Joyce Roberts' yard. He jumped several chain-link fences and vanished behind a chorus of barking dogs.
>
> "I ran in and called 911," said Charlotte Pearson, who was having the sale. "I thought it looked like a reindeer, but they said, 'you mean the elk.' "

From another report:

"When I glanced out the car window, I thought it was a cow or horse," said Gene Eyer of Independence. "I got the car stopped and got to looking close . . . there aren't too many horses with a hat post on top."

In April of 1990, the elk was in a wreck, hit by a car while it was crossing the road. Said one of the drivers involved in the wreck: "My first thought was that it was a big dog, because it had this white collar."

The animal had appeared so often in the press that reporters had taken to naming him. Eventually, he would be known as Montana, Big Guy, and Earle of Montana. This became truncated simply to Earl.

The elk escaped the wreck and capture, setting all of Independence and surrounding towns to believe he had been mortally injured by the collision and had simply gone off in the woods to die. He hadn't. In June, he was spotted again, and this time attempts to capture the animal ensued.

In September Earl was finally drugged and captured near a football stadium where he had been grazing. The collar ultimately led back to Olson, who identified the animal as the subject of his experiment three years before in the Sweet Grass Hills.

The elk was held at a zoo, where it kicked apart a gate and a one-thousand-pound bale of hay by way of protest. Ultimately, he was moved to a county park. Then Montana governor Stan Stephens, who headed an administration notorious for its disdain for the state's wildlife, told Missouri it could keep the elk with Montana's blessings.

"We know that #964 will be able to live a long and happy life at Jackson County's Fleming Park and we are thankful for that," wrote the governor.

I once spooked a young black bear that had come to my house's front porch to test the door. It whined at me, then shuffled up a nearby pine tree to fidget and cry. I ducked out of sight and it climbed down, then I huffed and snarled at it to scare it off into the woods. Once it was satisfied it was out of my range, it slowed its panicked dash to a normal bear ramble. I hid and watched it.

I saw in that bear an overwhelming sense of the Other, the essence of wildlife that so seizes our attention. One time I heard a native man quoted as saying, "Every animal knows way more than you do," and the bear seemed to say the same.

Its path away from me followed no course that made sense, no normal path of flight or of purpose. Rather it zigged and zagged up a hill, stopping here to eat a plant, there to eat a plant. Then it occurred to me the bear was eating not just plants, but a particular plant, one species out of the hundreds then present on my land. He was reaching into a bear's sense of place to read a sort of natural map of the landscape, and this map guided his travels. The map was written in vegetation. The bear's course was as purposeful as a politician's path through a roomful of potential donors. The bear was simply reading nature's directions for inhabiting its place, and those directions are manifested in plants.

It's not hard to imagine that a long-traveling elk could be pulled along by just such a set of directions, a map if you will. There is written on the land an inherent sense of the place. Because our culture has not learned to read these directions, we substitute our own vision, lines, an abstraction that we call a map. What, in the alternative, could we learn about the planet's instructions if we sought our vision by following, as one lone elk did, the planet's own instructions as they lead us out across the plains? Might we find what he found? Might we find the elk, or more important, might we find the sense of the elk?

3

What the Wind Carries

Again I travel, again east, over the Great Divide to the open of the plains. This trip comes at winter's end, so the plains look all the more boundless to eyes that have spent all of this northern latitude's dark months in a box of a house in the woods in the mountains.

I am driving an interstate highway hard, crossing the Divide near Butte, Montana, on a flawless February day. In the sweep of high plains, there are layers of grass cured to a soft yellow shine, a few streaks of ice and old snow, a wisp of dormant cottonwoods in the draws, pine on the slopes of the hills, and cloudless, clean sky overarching all. Even my beat-up old Volvo seems capable of flight.

I have read about and brooded on the plains the whole of the confined winter, yet it seems that in the last five miles of pure motion, I have learned much more about them. The big grassland begins just ahead of my grill. On my left it stretches north through most of Alberta and Saskatchewan, on my right south to Texas, and dead ahead to Minneapolis, which lies two hard days down this interstate. These first views of the plains from mountains often tempt one to try to absorb the whole in a gulp—impossible, but the effort pays. To try is to try to become the wind.

It becomes clearer the wind is our quarry. The wind is everything here on the Divide, a verity, a player easily missed. It's easy to think the fender-rumbling gusts are only a freak of the moment, not the rule of the place. The lean of the trees and the clean sweep of the land say otherwise. Confirmation comes again just down the road past Bozeman at Livingston, a quaint cowboy town in recent years filling fast with artists, writers, and movie stars appearing in their latest roles as ranchers. There are signs on the interstate announcing that occasionally the road will be closed to trucks towing trailers because winds can and do tip them over. To the south of the highway stands a patch of windmills, high-tech and modern, each the size of a radio station's broadcast tower. These are Livingston's experiment in catching the wind for electrical power. Half of these towers, though, lie flat on their backs like KO'd fighters, leveled by an ancient force.

My car and I are headed east—with the wind as it pries off this last range of hills, a runner off a starting block. With the wind and headed east—many here travel this way. Parallel to the highway, the Yellowstone River shakes off a winter's skin of ice and piles it in bends and sloughs. Once unburdened, the water runs east, then northeast to join the Missouri River. So the river, the car, and the wind all run parallel to the journey of that elk from the Sweet Grass Hills, parallel to other journeys that moved across this plain through the millennia. The story of the West is the accumulated record of millions of journeys. It is written in tracks in the dust and the snow. It is the story of the goings of creatures with no ability or inclination to leave a record. There is a record of sorts, but it's as ephemeral as the wind's. We may as well be chasing ghosts.

From the highways, it does appear as if the West is over-fond of chasing its ghosts, the impression one gets from the towns that spread along the interstate system that is the funnel for summer tourists. It appears this is a region terribly entertained by its past, to the point of preferring it to the present. This fascination, however, extends only to the narrowest sliver of past, as if all of this big place could be captured with what happened in the last half of the nineteenth century, a generation's time. As if this great, implacable plain could utter even a single sentence of significance in the geologic nothing that is the span of a lone human life.

My day's journey in February that started in the mountains of Montana ended by coincidence in the Black Hills town of Deadwood, South Dakota. The final chapter of the plains Indian Wars began near here in 1874 after General George Custer, responding to trespassing by miners in the Lakota's Black Hills, made a survey of the territory and announced the hills were filled with gold "from the grass roots down." This revelation quickly concluded that the Sioux no longer needed the hills that had been granted to them in 1851 by a formal treaty that had been reaffirmed only six years earlier. Custer evicted them, and Deadwood sprang as a mining town and gambling town. It still is both.

There's a main street lined with Old West storefronts, some of them original. The actor Kevin Costner is repaying his moral debt to the Sioux and to the area that was the setting of the film *Dances with Wolves* by building a new eighty-acre "destination resort" casino just outside of town. Most of Deadwood's existing casinos lay some sort of claim to having played a key role in the final day of Wild Bill Hickok. He was shot and killed in Deadwood while playing cards.

His real name was James Butler Hickok, but he earned the name "Duck Bill" from a protruding upper lip, which he attempted to cover with his trademark drooping mustache. Later, at his trial for the murder of a man he had killed more or less for the hell of it, Hickok told such lies as to earn the name "Wild Bill." Buffalo hide hunter, liar, and common killer, this small-town hood became legend courtesy of the yellow press of the day. Bill is the chief ghost of Deadwood, where today retirees in permanent press stalk nighttime's Main Street, aiming the white bills of their golf caps for the next row of slot machines.

At Saloon Number 10, the weeknight crowd seems composed

of equal parts vacationing schoolteachers and off-duty cops. A seasoned waitress in a peekaboo black lace dance hall get-up ferries a tray of wine coolers, margaritas, and Lite beers. Every customer is white.

The rough-sawn walls are layered in rusted muskets, horns, heads, hides, and hooves, antique photos of men in big hats and stiff suits, a velocoped, a grubstake gold miner's tools, various documents, and a requisite copy of the painting Anheuser-Busch commissioned, *Custer's Last Fight.* Next to the painting is a photo of the first aircraft to land in the Black Hills, thirty-eight years after Custer kicked out the Sioux.

A chair hangs over the door, reportedly the very seat that held Hickok's sorry ass when Jack McCall's pistol ended the game. Over the bar, death masks of Hickok, Calamity Jane, and McCall leer over the proceedings. The lounge band of aging hippies plays Elvis's "All Shook Up." This is the prevailing memory of the Old West, but there is West that is older.

Fittingly, it was a horse that kicked off one of the key intellectual battles of the nineteenth century, specifically a dead horse found near a train station in Nebraska. What we must first understand about the West lies encoded in the mouth of that horse. By the time it was found, modern horses were completing a revolution of plains life, begun by their importation.

Hernando Cortes brought sixteen horses from Spain when he sailed with six hundred men to conquer Mexico in 1519. Within 250 years, horses had spread to enrich the nomadic life of plains Indians. This, however, was not an introduction, but a reintroduction. Horses evolved in the New World, migrated across the Bering land bridge to the Old World, then became extinct in the new until the Spanish brought them back.

Our horse in Nebraska had been dead for sixty million years. It was one of the fossils pivotal to the work of the Yale paleontologist Othniel C. Marsh. By 1876, Marsh was using his collection of extinct plains horses to assemble an argument that overcame key objections to Darwin's theory of evolution. This achievement caused considerable stir in Europe and a petty, ill-mannered, and petulant tiff in this country between Marsh and his chief rival, Edward Cope, who challenged the veracity of Marsh's work. The two scientists chose up sides as in the schoolyard, dividing the nation's entire scientific community. The fallout from the battle was to ripple throughout the tail of the century, leading to court battles and charges of fraud and plagiarism, and reverberating all the way through the government's attempts to conduct its conquest of the West in a scientific fashion.

Marsh's diagram of horse evolution wound up in the science textbooks of the day, including that of the Tennessee teacher John Scopes, who was to be prosecuted by the great prairie populist William Jennings Bryan in the Scopes Monkey Trial. Cope's work and arguments eventually faded from view.

A single animal on the plains can create quite a stir, yet what we need to know from it just now has little to do with the stir, but lies resting in its mouth. The record of horses runs back sixty million years, but about twenty-five million years ago there occurred a development key to Marsh's description of evolution. Horses developed high-crowned teeth (taller teeth especially adapted for eating grass). At about the same time, so did camels and rhinos, also residents of what would become the Old West.

From this rising of teeth we can know the Rockies had risen a few hundred miles to the west, and from this we can know the age of grasses had begun. The creation of the American Great Plains as grassland was a function of the creation of the Rocky Mountain chain. Before, the region had been a vast sea, then a vast flat forest. The puny teeth of the residents said so. The Rockies, however, rose and raised their rain shadow, which produced the aridity that killed the trees and created the grassland.

More subtly, the rising Rockies demanded of the plains' plants a subterranean adaptation. The uplift of the mountains created a gentle tilt of the old sea bed, a consistent slope of ten feet to the mile from the mountains all the way to the Mississippi. This tilt meant the plains would drain, but slowly. The river system would then become a rack of east-running streams carved to gullies, then filled by sediment, and then carved to gullies again by seasonal runoff. These rivers would mostly carry water from the mountains, not local rains. The rains on the flats between the rivers would soak in so that plants there would evolve a complicated network of roots to exploit the moisture. This, then, became the sod, the secret society of prairie plants. Above ground, fire, winds, blizzards, and grazers could decimate the visible part of this plant life, but as long as sod survived, so would the society to hold the thin soils on the gentle grade.

Biologists use the term "biomass" to describe the weight of life of a place. Life in the tropics, for instance, manifests itself as forests, and there almost all of the weight of life is in the trees' trunks, almost none in the roots. That's why deforestation there is so devastating. In grassland, the vast majority of biomass rests in roots. They are the center of the life of the place, and all else flows from them. In grassland, above-ground attacks are tolerated, but destruction of the roots is destruction of the place.

Aridity begets grass, which begets high-crowned teeth; all of this the rising Rockies decreed. The wiry stalks of grass are less nutritious than the leafy fare of wetter regions. Therefore, grass eaters must eat more food and more abrasive food. To do so, they must have teeth to withstand the strain. For twenty-five million years, the plains have demanded and gotten a unique set of residents, a fact that has not changed.

My ultimate destination was not Deadwood. I was headed for the south edge of the Black Hills where they tail out into the plains. I had planned to finish my trip with a half day's drive through the hills, weaving the roads among the pines, tourist traps, and the place where an eccentric named Borglum decided the highest and best use of the limestone cliffs was to hold effigies of U.S. presidents. There had been snow the night before, however, and those winding roads were slick, so I chose to take the interstate that skirts the hills to Rapid City, then a prairie highway that cuts on south, to complete the loop.

Tractor trailers hogged the interstate, claiming wide berth on the fresh slush and ice. The FM band offered a variety of entertainments, any sort of country-western music one would want to hear. Weather forecasts were frequent and earnest, as if broadcast to people who cared deeply about weather. Likewise the livestock reports: feeders were up; bred heifers, up; weight cows unchanged. That day, there was to be a cutting horse competition at the winter stock show in town, and every motel was booked. The radio's news program centered on an interview with the trainer who had won the sheepdog competition the night before. Billboards advertised Tony Lama boots.

The snow thinned to rain at Rapid City, then back to snow again as I continued south to Hot Springs, where I stopped for lunch at the old hotel. Like the rest of the town's more imposing buildings, it was built of trunk-sized blocks of red spearfish shale. The owner was new to the business, having just brought his wife and kids to the prairie from one of California's more frenzied metropolises. He spoke of worries of riot and crime in the past tense, although the move was, he allowed, a bit of a culture shock for the kids, but then one could get Nintendo and videos here, too, and next year the kids just might get interested in hiking.

Mostly, though, he wanted to vent some spleen after I said I was in town to look at the old bones. The hotelier was angry because the Smithsonian Institution was carting off a mammoth find from farther north in the hills, and those bones belonged in the hills. I had to agree, although not in sympathy to commerce. The bones are elephants, dead elephants, and they would seem less remarkable in an East Coast museum, like a fossil in a fossil zoo. They seem altogether remarkable here on the plains, where you can squat and touch one, then look out to the rolls and coils of grassland and try to imagine how this could have been. Not elephants of the trees, but elephants with high-crowned teeth that ate five hundred pounds of grass each day.

The mammoth site at Hot Springs, South Dakota, is a new development, the result of accident. In 1974, a local developer, Phil Anderson, started work on a subdivision on the outskirts of Hot Springs. Toward that end, he commissioned George Hanson to bring his backhoe and dig holes. Hanson's holes held bones.

Hanson had a son who happened to be taking a geology course at a nearby state college in Chadron, Nebraska. The son thought he recognized the tooth as that of a Columbian mammoth, an extinct species of the plains. It was, so Anderson backed off his development plans long enough to let the town figure out what had been found. These bones, however, were not to be carted east. Interested people formed an independent corporation that bought the land from Anderson. In 1975, volunteers from the national group Earth Watch began excavation supervised by experts, including Larry Agenbroad, the geology instructor of Hanson's son.

The excavators dug up a window in time. When Hanson sunk his backhoe's teeth, he was digging atop a hill, but not so long ago the site had displayed the reverse profile. The mammoth site is an ancient sinkhole, formed 27,000 years ago by the collapse of an underlying layer of limestone. Beneath the limestone was hot water under sufficient pressure to well to near the top of the hole, forming a heated pond about the size of a basketball court. If we judge by the leavings of aquatic creatures in the hole, the water temperature was about 95 degrees Fahrenheit. Couple this with the fact that this occurred near the height of the last glaciation. The sheet of ice by then was probably somewhere near its extreme southern boundary, which was the Missouri River, which is only 186 miles from the site. It is not difficult to imagine how a pool of hot water might have formed some attraction for the animals.

Oddly, though, the attraction seems to have been limited to males. The walls of the sinkhole, spearfish shale, grew slick from the simmering water, forming what became an elephant trap. Mammoth society is assumed to be similar to that of modern elephants, a matriarchal society. Females and young wander together in large groups, while males, when not needed for breeding, sulk off alone to wander the plains. The sinkhole ensnared only males, forty-seven of them at last count. All but three are Columbian mammoths, a plains species that migrated to this continent from Siberia nearly two million years ago. Interestingly, the remaining three are woolly mammoths, a species more often associated with Siberia and the northern, glaciated reaches of this continent. The Hot Springs site is the only mammoth site found to contain both species, a link between two worlds.

The sinkhole was not a swirling pit of emaciated elephants bellowing in their death throes. The hole remained open for only about seven hundred years, meaning it claimed a mammoth on average once every fifteen years. Then it slowly filled with sediment eroded in from surrounding countryside. The sediment cooked and dried harder in the hole, so when later scourings of erosion planed the surrounding plains, the harder mass left the sinkhole standing as a knob.

Today it is covered by a modern free-span building with a large foyer for visitors, displays, dioramas, and a book and souvenir shop complete with mammoth shot glasses. All this is forgotten soon enough, though, in the presence of Sinbad, a Columbian mammoth that looms over the foyer. Sinbad actually is a fiberglass model, with each model bone formed with a cast from an actual bone.

Columbia mammoths are bigger than modern elephants, standing about fourteen feet at the head. (The smaller woolly mammoth was about nine feet tall.) The longest tusk at the site was seven feet ten inches. The animals weighed ten tons and wore a sparse coat of hair.

They had four teeth at a time, each about the size of a fist. As front teeth wore down from the business of grinding up five hundred pounds of grass a day, the back teeth would move forward as if on a conveyor belt, and new teeth would sprout in the gap at the rear. Mammoths lived to be about sixty-five years old, growing six sets of teeth and perhaps dying simply because their teeth wore out. Eighty-two percent of the mammoths snared at Hot Springs were young to middling, fifteen to twenty-eight years old.

Most of them still are there in the hole. Sinbad, reassembled in fiberglass in the foyer, is the exception, but on passing through the foyer to the gym-sized roof over the dig, one discovers most of his colleagues remain in situ. Diggers have brushed off the piles of bone, leaving a skeletal tangle to reveal the ages, femurs as big as fence posts, pelvises like automobile grilles, and skulls like a field of boulders. One individual lies in a head-first pile of bones, a death pose that earned him the name "Acapulco Diver." Another is spread out a bit: "Napoleon Bone Apart."

Near an end of the hole, probably once a shallow end where there was a ledge, lies a smaller skull, sleek and more menacing than the round-lobed heads of the mammoths. It was a short-faced bear, larger—two feet taller at the shoulder, three feet longer, and more wiry—than the modern grizzly. Like the surviving bear, this animal was probably an opportunist. It dropped in to feed on a failing elephant, probably was injured, and died here.

There are hints of other animals in the dirt, mostly teeth that apparently washed in with sediment, teeth of a camel, a peccary, wolf, coyote, minks, voles, a deer mouse, prairie dogs, rabbits, pocket gophers, and ground squirrels. The pollen shows the plants hereabout were mostly grasses, with a smattering of trees and brush; this was a realm of vegetation and small animals not much different from what exists now.

The impression from the hole, with its twist of conglomerated bones, is one of a tortured world, yet this impression is inaccurate. Most of the twisting and torturing of the bones came long after the animals died, with the heaves of geology's settling sediments. The hole is not the leaving of a sort of cookie cutter that punched us out a cross section of late Pleistocene life. Rather, it is a piece of highly selective flypaper that managed to arrest the motion of some who passed nearby. These animals died, but so many others endured, and that, at bottom, is the question for us and for the plains. We must know what endures.

The elephants are not at all irrelevant to this question, given the context. In fact the clues suggest the big animals are the only beings we would find strange here in an otherwise familiar world, so the question rolls back on itself. For two million years, elephants endured in a world that was, in geologic terms, in a state of constant upheaval. Why did they suddenly disappear, not 23,000 years ago—the last years of this hole—but last another 12,000 years to practically yesterday, to almost precisely the time of the last retreat of the glaciers, then suddenly die off en masse maybe 11,500 years ago, much later still in the case of the woolly mammoth in Alaska? Practically yesterday.

Elsewhere around the continent, in sites as widespread as Wyoming, Michigan, Arizona, and New Mexico, there lie the skeletons of more recent mammoths, some of the bones smashed, cut, and flaked, some of them bearing clear marks of artifice, some found near distinctive stone spear points we call Clovis. The world of North American mammoths is today available only by reconstruction in our imagination, but for some humans, residents of our continental grasslands, this world was real. For a short time, humans and mammoths coexisted in the New World.

The question of what endures in grassland breaks down quickly as one travels, finding so many islands of what is so different among what appears so much the same. One can walk what at first seems a limitless expanse of grass, only to drop into a draw and find forest. One can climb a hill in Montana only to find plants otherwise at home only in the Arctic Circle or the desert Southwest.

First arose the mountains to the west and then came the dry winds that made aridity that made grassland, and the grass made the animals that go with it, the horse, camel, mammoth, bison, and a circle of co-dependents.

Then, nearly two million years ago, began an era we call the Pleistocene, the Ice Age, and all of the northern reaches of the continent became periodically crushed beneath a great elastic sheet of ice, but not all at once, and not of a sudden, and that's what counted. The glaciations came in periods governed by a phenomenon called Milankovitch cycles, after the astronomer who discovered them. Periodically, the earth's orbit becomes more elliptical, increasing its distance from the sun. Each of the hundred-thousand-year-long cycles has been marked by sixty to ninety thousand years of glaciation followed by forty to ten thousand years of warmer periods or interglacials. Each glaciation built slowly from the north as increasing snow cover reflected increasing amounts of the sun's already diminished heat. The ice piled deeper in the Arctic, squeezing and pushing itself south, some years retreating, some advancing, but on average advancing, until a glacial maximum was reached toward the end of each glaciation.

Against this advance, the living world retreated. The zone at the edge of the ice carried a moving ring of boreal forest and tundra, sometimes mixed with grass. Farther south, dryland plants made room for migrating species from the north. The animals presumably followed their foods of choice. Always there were winds, enormous winds sweeping off the unobstructed plain of ice. Wind and cold stripped vegetation, leaving soil exposed, creating dust. This dust piled in dunes that became the foundation of grassland soils today, a wind-bred soil ready to run again at the slightest excuse of wind.

The interglacials probably were periods altogether similar to today's climate, so similar that there is no good reason to believe the ice age has ended. The cycles that caused the glaciers remain. Our era is somewhat arbitrarily labeled the Holocene, as distinct from the Pleistocene, but really it is nothing but another interglacial. It has already lasted the requisite ten thousand years and may be ready to end, or maybe not. Previous glaciations built after the global climate cooled only a couple of degrees over a couple thousand years. Global warming triggered by industry will warm the earth that much in less than a century, so all bets are off.

Massive climatic change has been the rule of the grasslands, and this is a reality not limited to the periods of glaciations. The record since the glaciers retreated, even the record of this century, shows motion. About a thousand years ago, the globe entered a sort of mini–ice age, well documented in the written record of Europe. In mid-America, the cooler temperatures brought extra rains. An archaeological site in Iowa's tall grass prairie produced the bones of forest creatures in layers dated to nine hundred years ago. During wet years, the trees advance. A few hundred years before, all the bones had been from bison—grassland animals.

About every thirty years, the arid grasslands become more arid still, in the regular droughts that farmers call catastrophe but that are actually normal. During these cycles, the border between tall grass prairie and mixed grass shifts as much as a few hundred miles east. Plants get shorter. Some disappear altogether, into their long-lived roots. Some disappear to long-lived seeds and lie dormant in the dry ground. Always there is motion, retreat and advance. The record is clear: What endures on the grassland is motion. In the long view, in the short view, on micro and macro site, the grassland is a place of motion. It is a place of upheaval, and one deals with upheaval simply by getting out of the way.

Almost all of the plains Indian cultures divide the universe into the four directions; a salute of these four is as ubiquitous as the sign of the cross in Mediterranean Europe. North, south, east, and west, and even, among some, up and down, for six directions. North and south in the face of the advancing ice every one hundred thousand years or the advancing cold every fall. East and west away from and into the rain, upstream, downstream, into and with the wind. And always up and down. Up slope to ridge top, where it is wetter, or to hilltop out of the ice sheet and in refugia. Up slope and down slope of draws, down to the creek in dry times, away as water rises. Above ground and tall in wet years. Below ground and to the roots in dry, or to the seed to ride on the wind. Living things move to avoid the extremes.

In all, the West is a place of journeys. It is a place so attuned to motion as to require it for its survival. It seems, then, the richest of ironies that the two successive waves of upheaval to this world, the most profound unsettling of this world, began in the journeys of two very small bands of animals, separated by eleven thousand years.

At the time of the last retreat of the glaciers, all of the American elephants died, and they were not alone. At the same time 73 percent of the large mammals of North America became extinct, including thirty-nine genera of mammals weighing more than a hundred pounds. Gone were a couple of species of peccary, a camel, and a sloth, the horse, two species of llama, an indigenous moose and two species of deer, one of antelope, a couple of musk oxen species, the mastodont and mammoth, saber-toothed tigers, a giant beaver and a giant capybara, a tapir, the spectacled bear along with the short-faced bear, a cheetah, and armadillo.

This had been a stable world, if such a word might be applied to the plains, and that is just the point. It was a world that had endured intact the comings and goings of glaciers, drought, and cold for literally millions of years, only to disappear all of a sudden 11,500 years ago. The animals that disappeared had already cut their peace with upheaval, so what was new?

There have been waves of extinctions throughout the vast history of the globe, which spread worldwide, but this one was largely limited to North and, to a lesser extent, South America. (Australia and New Zealand had undergone similar extinctions shortly before this, a further anomaly that becomes another piece of the puzzle.)

Further, unlike earlier extinctions, this devastation was for the most part limited to large mammals. Plants fared well; the grasses that fed the animals endured. Small mammals, birds, and invertebrates suffered almost not at all. Likewise, there was no sweep of death in seas or nearby aquatic habitats.

Some large mammals did survive in North America, but an examination of those lends to the puzzle. Of all the animals we call native today, the bison, elk, deer, moose, grizzly bear, black bear, caribou, and antelope, only the antelope, having evolved on this continent, really is native. The rest are Eurasian. They came here across the Bering Strait with man, just yesterday in the geologic sense. These invaders survived, most of the rest are dead, and we should know why.

Paul S. Martin is a tall man with a warm, patrician face, a soft fold of gray hair, and eyes that suggest there is always something on his mind. He is a vegetarian and walks with aluminum crutches, his legs twisted in. On the pleasant spring night we talked on the patio of his house in an adobe neighborhood of Tucson, Arizona, he wore a cheap pair of corduroys, a plaid western shirt, suspenders, and Birkenstock sandals over socks, the same outfit he had worn the day before at a lecture at the University of Arizona, where he is professor emeritus of earth sciences. In retirement, he still uses his cloister of a lab, where he spent every bit of a career using his stilt fingers to sift through "sloth shit," his term. Among paleontologists, he is regarded as a master of the dung of the extinct ground sloths that populated North America, part of the range of animals known as the Rancho Labrean fauna, after the hole in Los Angeles where the remains of this menagerie of large extinct mammals were first discovered.

Martin is not known so much for his knowledge of dung (called coprolites in polite circles and sifted for what it contains of plant remains, therefore providing a window into a lost world) as for a radical and unsettling idea he has held for close to forty years. Martin is the chief proponent of the notion that the wave of extinctions of large mammals at the end of the Ice Age was caused by human hunting, a notion he calls the blitzkrieg.

On the patio he speaks of this in measured, quiet tones punctuated with frequent smiles, the debating habit learned from a career spent defending an outrageous idea. Listening quietly and politely with me is Andrei V. Sher, senior scientist with the Severtsov Institute of Evolutionary Morphology and Ecology of Russia's Academy of Sciences. Sher is one of Russia's leading experts on the Ice Age, particularly mammoths. The day after our conversation with Martin, Sher would make headlines around the world with an announcement that a dwarf of the woolly mammoth survived until only four thousand years ago on Wrangle Island in the Bering Strait. On the day before this announcement broke, Sher was bouncing around southern Arizona in a Chevy Suburban, visiting a couple of famous holes that Martin had worked to assemble an indictment of our species.

Sher, a small man with a black beard and thick glasses, simply drank his beer and munched chips and salsa while Martin expounded on an idea with which Sher does not agree. Sher is a chief proponent of the other school of thought, that the climatic upheaval that ended the last glaciation about twelve thousand years ago was the beginning of the end of the mammoths and the rest of the Rancho Labrean fauna. Maybe all of this should come down to a quiet conversation with chips and salsa, because between the two men the range of knowledge of this matter is encyclopedic and global, and still they know anything they say will not resolve the issue. The debate has raged since Martin first proposed the blitzkrieg in a paper in 1959, and still nothing is resolved.

Over the years, though, Martin's idea at least has risen from the status of apostasy to one of two dominant ideas about the Ice Age extinctions. Some say it has become the dominant idea, the truth of the matter. In his watershed 1992 book, *The Diversity of Life,* the renowned biologist E. O. Wilson signed on firmly with Martin's camp.

Martin's argument has been shored and enlivened as the holes and dung heaps around the continent offer evidence, shored until there is no danger of the idea's going away anytime soon. Now the two ideas—hunting and climate—have been opposing each other so long that one begins to suspect they will someday achieve the status of all great dichotomies, that one will never prevail over the other, but one will explain the other in a synthesis, just as Martin, in conversation, can help explain Sher.

Whatever the case, we need Martin's idea just now for our story.

What it tells of humans is intimately wrapped in our journey and in grass.

Some three million years ago the species we have come to call human was nothing but an undistinguished member of a range of small forest apes. Though much smaller, in abilities and physiognomy this ancestor was not that much different from us. Theoretically at least this *Homo erectus* had become capable of those feats that the smug among us think earn us the title of crown of creation.

The little ape, however, was not particularly successful, and evidence suggests it simply eked out an anonymous existence in a corner of the forest. Then the rules changed. A shift of climate drastically shrank the forests of Africa, creating the great sub-Saharan grasslands. It turned out the little ape was ideally suited for the new terrain. Lacking a tail and the agility necessary for treetop life, humans were disadvantaged in the forests, but with the ability to stand erect and see over the grass, humans fit grassland.

Always I believe that the rush of freedom I feel on encountering an open vista of grasslands is racial memory of a time when our species learned to run among those first wide open spaces. Indeed, there is a hypothesis called "biophilia" that suggests this very thing, that knowledge of nature brings a love of nature, and both were necessary adaptations for survival by primitive humans. We relied on a deep knowledge of our environment as much as we relied on our opposable thumb. Because humanity's deepest memories were formed in grassland, this is where our deepest knowledge and love lie. To this day, we feel at home on a bluff overlooking a meadow, which is why New York City high-rises overlooking Central Park bring the highest prices. The preference for a room with a view is a grassland ape's relief at being freed from the confines of the forest that hid her enemies.

Over the millennia, the erect ape learned to exploit the grass, the big animals that roamed it, and the community of plants that carpeted the drylands floor. Finally, with tools and spearpoints, we spread north and out of Africa. Family lines like Neanderthal and associated technologies arose, then died, probably not extinct but married into the main line, the bumps smoothed off until we became simply *Homo sapiens.* When someone suggests that Neanderthal man was simply a crude forebear of the refined creature we consider ourselves, I think of the forty-thousand-year-old grave site in Iraq where a Neanderthal is buried with a sprig of flowers. These people are us, the same creatures that would go on to domesticate horses, dogs, grain, and hydrocarbons, build libraries, temples, and slums.

Our ancestors did eventually learn to thrive in forests, moving north to the boreal fringe in Europe and Asia. But our firmest tracks, the earliest paths, ran through grass, out of Africa and north to the Mideast, where we would first cultivate the grass we now know as wheat. On we went to the broad band of grass that is the steppes of Asia, to hunt elephants, the woolly mammoth, and a host of Eurasian grazers. Here arose one of the globe's most firmly entrenched traditions of hunter-gatherer nomads.

In Paleolithic culture, the Stone Age, we chipped and fluted points and scrapers to make hunters' tools. In a small corner of northeast Asia, up near Siberia, one group of our people became genetically distinct, coming to possess a tooth hollowed like a shovel, a characteristic known as sinodonty. This signature tooth would become the mark to trace a population explosion in the New World.

School kids learn to call it the land bridge, a metaphor both wrong and right. The broad, flat plain that linked Alaska and Siberia across the Bering Strait at the height of the last glaciation was too imposing to stand in mind only as a bridge. This land paleontologists call Beringia was a place in its own right.

On the other hand, it was indeed a bridge. It carried a collection of big mammals with high-crowned teeth and an upright, shovel-toothed primate that owned a technology that would bring revolution to a virgin land. Maybe as early as forty thousand years ago, but almost certainly by twelve thousand years ago, a tidal wave of upheaval had begun.

Beringia was not simply the strip of land that connected the continents; it included the connected coastal plains of both Siberia and Alaska. It was ice free in an otherwise iced-over world. The same glaciers that had tied up the earth's water and so lowered sea levels to expose the land bridge had reduced precipitation to produce an arid climate. It was plenty cold enough to snow, but there was not enough atmospheric moisture in the vicinity to allow it.

Beringia was flat, arid, and cold, a grassland. The plant record is clear on this from both Siberia and Alaska, from the layers of deposits through the millennia. Most recently these were the leavings of the taiga—the larch and birch forests that cover the area now—but the trees fade from the record as it works its way back to the Pleistocene until eight to nine thousand years ago, when the record shows a land of grass and sagebrush. It was a land bounded on both sides by glaciers, but it held its own population of grazers, including the woolly mammoths, camels, bison, and horses—from a human point of view, meat. Accordingly, these animals drew the stone spears and scrapers of humans.

Andrei Sher, whose life's thought centers on this place Beringia, says the grassland there, cold as it was, is without modern analog. When the global climate warmed, this unique plant community retreated farther and farther north, to what is now Siberia, followed by the mammoths, which were eventually rendered extinct by the loss of the plants that made them.

"You want to know what caused the extinction of the woolly mammoths?" says Sher, showing a slide of a sodden modern tundra. "I say lakes and moss."

And that may be, but the camels, horses, and bison survived, then thrived in grasslands of the Old and for a bit the New World, global change or no. Further, the Columbia mammoth, a larger elephant more adapted to the warmer and arid reaches farther south, died out in the New World, although analogs of its grassland abound. There seems more afoot than climatic change, in that something big happened and happened fast in the New World coincident with the retreat of the ice and arrival of what we have come to call Clovis man.

Viewed from the top, the modern globe presents a broad streak, most often color-coded dust yellow, which suggests the lineage of the modern American grassland. The band runs huge and strong from Eastern Europe all across Asia, finally tapering near the Bering Strait, but if our mind's eye fills in the gap with Beringia, also a grassland, one can see the band now joined just south of Alaska to the beginning of the American steppe in northern Canada. Properly considered, all of this is our continent's motherland, the whole of us. Divided by a sea, then balkanized by politics, then colonized and exploited by the surrounding tree cultures—the cultures of first China, then Rome, then modern Europe—this once whole land was the place that gave us wheat, corn, dogs, horses and horsemen, camels, and endless wandering.

During the Ice Age, the northern piece of the mother grassland in the New World was glaciated south to just below what is now the U.S.-Canadian border. But at least toward the end of glaciation, maybe for all of glaciation, there ran an ice-free corridor from Beringia south. It ran just west of the Rocky Mountains into ice-free land. It seems reasonable and now clearer from the archaeological record that this corridor was Clovis man's path to a world of unparalleled ease. The ease is assumed by the lightning speed with which Clovis man populated the New World and the condition of the bones that is the record of his kills.

Curiously, there was until recently a glaring gap in the archaeological record. In fact, the record is largely a great collection of gaps, more like a net than a solid sheet, and, in Martin's mind, this fact is a clue in itself. Europe is rife with ancient sites containing both human and mammoth remains, showing even an architecture based on mammoth bones. In the early going, archaeologists expected to find much the same in North America, but as evidence mounted during the last half century, it became clear Clovis sites were sparse and transitory in the New World—exactly, says Martin, what one would expect in the case of a fast-moving, colonizing culture, in the case of a blitzkrieg. The mammoths simply disappeared before Clovis began building houses with their bones.

Still, even in this light, there remained a disturbing gap. Most of the Clovis sites had been unearthed in the arid grasslands of the continent, especially in the Southwest, as well as Wyoming and South Dakota. None had been found in Beringia, the site of the colonists' beachhead. But in 1993, archaeologists announced conclusive dating of Clovis remains in northern Alaska that placed the culture there at 11,700 years ago.

There are claims of earlier human habitation on the continent below Beringia. One site in Chile has been set at twelve thousand years old, while there is evidence of human habitation in Pennsylvania nineteen thousand years ago. There is among scientists, however, considerable skepticism about these claims, and the dating is held to be unreliable. The Alaska site, on the other hand, falls nicely in the range of early Clovis time, a period of about a thousand years that encompasses the beginning of the retreat of the glaciers.

In archaeology, imagining is a dangerous business in that our speculations are channeled by the specifics of modern life. Yet in this deep-time travel, imagination is almost all we have. We cannot avoid speculating on the life of the first of these people, the shovel-toothed individuals who occupied the site in Alaska. In the stone spear in their hands they carried an efficient bit of technology that was allowing the spread of this primate from its original home range in Africa to global occupation, an achievement unique among species of big mammals.

Joining these people in the journey across Beringia were animals they had hunted for thousands of years, the swift, the secretive, the fecund, and the wary, animals familiar with the technology of spears. Before them lay a great gulf of grassland stocked full of naive grazing beasts who knew nothing of spears. Power can be measured only in relation to the power of others, and in this sense those first colonists in Alaska were people of enormous power.

That day in Arizona, the Russian scientist Sher was visiting for the first time land famous in his field. A group of University of Arizona scientists loaded him in the Chevy at Tucson early that spring morning for the two-hour drive south out of the sprawling adobe city. Tucson proper sits squarely in the Sonoran desert, a landscape of saguaro cactus, creosote bush, and ocotillo. It sits at the edge, though, of the great mid-continental grassland where it slips around the southern tip of the Rockies.

As the Chevy crested a stoop of a hill into Sonita Valley, the landscape instantly shifted from cactus to miles of rolling grass hills, sparred here and there with oak. This was among the first places in the United States claimed by Europeans. Coronado is thought to have visited in 1540. A Mormon battalion explored here in 1846, fleeing after one of its members was attacked and killed by feral cattle. The Apache Wars played out all around, and Geronimo surrendered in the nearby mountains in 1886. The gunfight at the OK Corral featuring Wyatt Earp et al. occurred just up the road at Tombstone, now a tourist trap. During the Mexican Revolution, various partisans took to mixing it up at the nearby border town of Naco. Aircraft under the command of Pancho Villa became confused in battle, strayed across the border, and dropped bombs, making Naco the only U.S. city ever bombed in an act of war. Now, a shiny dirigible, a radar installation meant to detect international trade in drugs, hovers perpetually over the army town of Fort Huachuca.

"You can see why the Spanish liked this area. When you get into this grassland, it looks like parts of Spain," says C. Vance Haynes, the University of Arizona earth scientist who has spent a full career sifting the valley's remains. Haynes knows the landscape layer by layer, epoch by epoch. He has been a full-time student of the earth since the day in the 1950s when he was stationed in the air force in New Mexico and found a mammoth tusk and tooth in a sand blowout.

"I look at this as mammoth country," he says of the surrounding plain. "This is pretty good mammoth country."

We travel another hour, over a squat divide and into the valley of the San Pedro River to a dry wash, one of the most famous in the world but marked only by a lone placard as the Murray site. A few miles away lies the Lehner site, so named because it lies on the Lehner family ranch. Each has produced hundreds of bones, the remains of at least twenty-six bison, camels, horses, and mammoths, thousands of the distinctive flakes that were the spear points of Clovis people, and charcoal from their cooking fires. All of this dates to 10,900 years ago, through radiocarbon dating that has been cross-checked with a variety of methods.

One mammoth rib cage holds eight spear points. One mammoth carcass lies at the head of a line of preserved tracks across the stream, perhaps its own tracks. Near the carcass there is a random shuffle of mammoth tracks, mostly large, in an arrangement caused by a habit observed in modern elephants: They are drawn to the carcass of a family member to pace and fret. There are smaller depressions mixed in with the big tracks, maybe human, though they're too fuzzy to tell. Across one large track there once lay a haft-straightening wrench, a stone tool that looks like a giant needle with an oversized eye. It was broken in half, as if stepped on by an elephant, and that's probably what happened.

This carcass was left largely intact, although two limbs had been "disarticulated." The bones from one of the legs showed up nearby, as if Clovis had killed the whole animal to eat one leg, the habit of a people used to plenty.

Despite all this evidence of slaughter, Haynes is not as nearly convinced as Martin that the late Pleistocene extinctions were solely the result of killing. He reads the stream-cut's walls near the bones, a slice in the earth maybe ten feet deep that is not so much a stretch of space as it is, to a geologist's eye, a stretch of time. The layers of tawny and red southwestern mudstones, marl, and sand lie horizontally, clean and sharp as the lines of a bar graph, thirty thousand years of testimony in depositions.

Haynes's thinking rests on a sometimes pencil-thin line of black that runs between the middle layers of strata, a structure he calls the "black mat." It was caused by deposits of algae after water flooded this area. Just below this lies the record of a long, severe drought. It is this black line that contains the site's crucial remains, and so Haynes believes the mammals and their hunters lived here at the end of a drought. This now-dry wash was a water hole during the drought, drawing in the thirsty elephants and making them easy prey for Clovis points. Therefore, climate had a hand in these killings.

But then these grazers had known drought through the millennia of their history and had survived. Can their demise be the result of climate alone?

Sometimes the black line along the stream bank is the width of a hand. Yet the stone-cut bones and spear points always rest at its very bottom, even when it is thick. Just a thin razor of line in all this recorded time means something happened here very quickly. The stratigraphy announces a stiff fact: The spear points appear suddenly, and above them there are no bones of Rancho Labrean animals. This is time, not space; after, not above. There are no native megafauna after the spears. No mammoths, no sloths, horses, camels, dire wolf, no *Bison antiquus*. The thin black line is the demarcation of extinction. The spears came and the animals went.

Wrapped up in Paul Martin's notion of the blitzkrieg is the idea of poverty, the word that best frames the dichotomy that channeled his thinking for forty years. It came to him first when he was doing some post-graduate work at Yale, between his time at the University of Michigan and the job at Arizona, the only job he would ever hold. He first began pondering the extinction of moas in New Zealand, a relatively recent event that coincided exactly with the arrival of humans there in the last one thousand years. Martin's work then took him to Mexico, where he first noticed a hole in the range of animals.

"There was a curiosity in Mexico. All the birds down there are richer than here. There are more plants. More birds. Yet the big animals are the same or fewer," he says. "Why are big animals impoverished when everything else is speciose in Mexico?"

That led Martin to an inventory of recently extinct species, what he calls a sort of crude checklist. From the checklist emerged the conclusion that, at least during the last million years, most of the extinctions are clustered around the end of the Pleistocene, about twelve thousand years ago, and most affected only large mammals, not birds, plants, or invertebrates. Further, the extinctions came at various times around the globe, coincident with the spread of humans.

The case in North America is particularly acute. In the continent's three million years prior to the end of the Pleistocene, a total of twenty genera of large mammals became extinct. In the space of a few hundred years at the end of the last glaciation, a blink of the eye by comparison, thirty-three genera ceased to exist, including all camels and horses, animals that had evolved in the caprices of American grasslands.

Equally telling, though, is the inventory of the survivors. The continent now holds twelve genera of large mammals that are, with one exception, all invaders from Eurasia. The moose, elk, caribou, musk ox, grizzly, and deer came across the land bridge with man. Even the species of bison we came to associate with the plains was an invader, replacing *Bison antiquus,* a straight-horned behemoth whose fossil record ends abruptly with the arrival of spear points. The closest living relative of the American bison, some say a subspecies, lives not on this continent, but in Poland.

All of these invaders that had survived thousands of years of hunting by humans and so presumably had evolved survival techniques, were animals Martin calls "gracile," in contrast to the sluggish sloths and mammoths. The lone surviving native is the instructive exception. The pronghorn antelope of the high plains is the continent's fastest mammal, capable of speeds of up to seventy miles per hour, not the best target for a spear.

Martin's first paper on his idea, written in 1959, was filled, he says, with "wonderful things that were incredibly wrong." The core of the idea has held, however, most recently confirmed by a computer model that shows the rapid spread of humans and decimation of animals to be mathematically possible in the short time in which he thinks it occurred. To be sure, the notion is not universally accepted among paleontologists—but then, almost nothing is universally accepted among paleontologists.

"We're trained to deal with complexity, and we're trained to be very suspicious of simple solutions. It's sort of if the priesthood had a bunch of parishioners that could see the nature of creation without anybody else's help. It takes away the opportunity for employment," he says. "There haven't been any benefits yet to the idea of blitzkrieg other than making a good story. People who are interested in simple answers—in history and economics and ecology and poetry—they are drawn to it, and people who work with the day-to-day evidence in archaeology and paleontology are put off by it."

Yet even if overkill is not the sole factor at work in the wave of extinctions, it seems to be an idea we need just now. It is humbling. It is frightening in its implications for our ignorant application of power. It expands our notion of history. If stone-pointed spears in the hands of Clovis were a technology sufficiently powerful to extinguish a band of species, what dangers adhere to our tools of today?

"The education of this country turns on more than Columbus. It starts with Clovis, and maybe we ought to start with a whole range of our patrimony based on forty thousand years," Martin says. "We structure our view of nature through historically what we see in America, and it comes down to home on the range and the buffalo and the deer and the antelope play and that's our image of nature. That's wrong. That's ecologically blighted in a serious way. It's a history that's confining our view of our patrimony, which has to include the mammoths and the mastodonts and the horses and the camels and the other thirty genera that became extinct. If those had been alive when Columbus arrived and we had seen their extinctions, think of what this would have done to the conservation movement. If the blood of dead ground sloths was on the hands of the European invaders in America . . . I think we'd be way down the pike in terms of preservation. We have to redefine our history."

Consider the wind of the West as inspiration, as that word lies tangled linguistically with wind, ghost, spirit, and breath. Understand that the wind carries ghosts that we should know. Understand that it carries the whiff of dead elephants.

4

A Lasting Peace

What is right in an ecological sense is what endures and what contributes to the endurance of the whole. Ecology's standard is conservative. The understanding that dictates this rule is co-evolution. The poet Robinson Jeffers said, "What but the wolf's tooth whittled so fine/the fleet limbs of the antelope," and that is the heart of the matter.

The living community exists within context of conditions: climate, soils, topography, heritage, and the community itself. In response to those conditions, the living members evolve a particular set of skills geared to survival, skills that are necessarily a reaction to existing conditions, including other organisms. If there are wolves, the ungulates will learn to evade them in sufficient numbers to carry on their line. If there is fire, the trees will learn the art of thick bark and sparse growth, the grasses the trick of retreating into their roots. Otherwise there would be no trees or grass. If there is water, there will be gills. If there is drought or cold, there will be migration.

Take away the wolves, and the assets of their prey—say,

fecundity—become a liability, manifested in overpopulation. Take away fire and the grasses and trees choke one another to death. In earlier times, lulled by the romantic literature of nature, we called this pitting of life forces against each other a balance. Then we got the long look, and we saw the record of the extinctions, and we saw the deserts and the arid grasslands with their sweep of drought, fire, boom, and bust, and the word *balance* seemed wholly out of place.

These are places of catastrophe to the point that catastrophe is itself a condition, a member of the community like any other. Removing it would be as wrong as removing wolves. We can only say that if a set of conditions has existed for thousands of years, then life has adapted to those conditions—drought, wind, grazing, fire, and hunting—and that any attempt to remove any one of them will produce stiff consequences and a new round of adaptation. There is an equilibrium of sorts, but it is a dynamic equilibrium. The planet accepts stasis only on the short term, our terms. The long term, however, is ruled by perturbation, creativity, and change.

From ecologists, one often hears the phrase "since the Ice Age," in establishing the legitimacy of a life form or force. This is, in temperate places, a rough definition of the time of the status quo. If a force has existed here since the retreat of the glaciers, then it belongs, it is right in the sense that all other life has adapted to it. It has existed here long enough to undergo the initiation rite of coevolution.

Given this, transportation—the process by which alien forms enter settled environments—and changes in technology—a quantum leap in skills available to a given life form—become terribly important. At the end of the last glaciation, the Bering land bridge was the mode of transportation, the stone-tip spear the alien technology. These came and all hell broke loose, but this was the Ice Age and what was then is gone. The intrusion of Stone Age men on the continents was then the upheaval but has since become the status quo. The issue before us now is what has happened since.

• •

The Clovis people and the animals that came with them across the land bridge, each bearing an alien technology of its own, had coevolved. There is some speculation that the human hunters had by then developed hunting rules and customs designed to husband the herds, but the rules did not apply to the new animals they encountered. They hunted these to extinction. Or maybe the exotic creatures other than people played a substantial role—say, the new bison or elk beat out the mammoth for grazing range, or the native short-faced bear found the new primate prey attractive, but did not suspect anything so advanced as a stone spear, so man became both bait and trap. Whatever occurred in the mix of climate, exotic incursions, and appetite, there clearly was upheaval.

Then a new order was built. To the extent we are tempted to level accusation at Clovis, we need to remember his age of destruction was an instant, followed by a lasting peace of isolation and coevolutionary change. There was a wave of extinctions 11,500 years ago that lasted a few hundred years, but then came relative calm of more than ten thousand years, during which there were no extinctions. This peace lasted until just more than one hundred years ago, when a new sort of transportation—the railroad—and the leap in technology that was industrialism screeched into the American plains, bringing not just death—death is always with us, a part of the status quo—but upheaval, by which we mean new waves of extinction.

Because Clovis traveled to the continent across a grassland, following the herds of bison and elk, it is safe to label him a creature of the grass. Once he was on the continent, the grazers and the wide grassland spaces drew him south through the land and probably produced an explosion of population.

Clovis, however, was not to stay Clovis, but eventually spun the growing numbers to new communities. Like his relatives in the Middle East, people who were at almost exactly the same time learning to weave linen and domesticate grasses such as wheat and barley, Clovis produced a settled, agricultural branch of the family in Mexico and Central America. And like his relatives, these people would domesticate a grass—corn—and as a consequence live in cities, learn writing, religion, and the autocracy that necessarily flows from irrigated agriculture. Just less than two thousand years ago, central Mexico raised the city of Teotihuacán, 200,000 people who survived on a network of trade and agriculture. At the time, the world held only one larger city, Constantinople, based on the domesticated grass wheat. Meanwhile, China was growing cities based on yet another domesticated grass, rice.

Clovis would also parent fish people, the salmon tribes of the Northwest, mound builders, traders, farmers of the Midwest, and the woodland hunters and farmers of the East. After Clovis, there were, as nearly as we can say, two other invasions by Asians before European settlement. One group with a distinct language and tool kit spread into the interior of Alaska to become the Athabascans. Their offshoots migrated at one point to the Southwest to become the Apache and Navajo, linguistically and culturally distinct from the Clovis-derived groups already there. In a separate wave about five thousand years ago, the maritime Eskimo people, the Inuit, spread across Alaska and around the Arctic Circle to contact European settlers in Greenland. Each of these groups produced a cultural strain worth following for what it must teach us, but our particular bias here is for the grasslands, so we will follow instead that strain of Clovis that was to become the buffalo people of the plains.

By 10,500 years ago, an Arctic version of the bison had sufficiently mixed it up with *Bison antiquus* to produce the new, smaller species finely adapted to the short grass of the plains: the modern American bison. The productive mid-stripe of the continent that had once supported herds of elephants now held the great herds of bison. There is no reason to believe that the gatherings then did not approach the massive numbers reported much later when white settlers began counting off clots of bison in the hundreds of thousands, even millions. These herds took days to pass, and when viewed from atop a tall butte, carpeted every yard of grass to the edge of sky in all directions. This was the basis of an economy of a settled West, a land that would remain settled for ten thousand years. It was an economy our history does not recognize as civilization simply because it was not based on a domesticated grass and on literature but rather on wild grass, meat, and stories that moved with the people.

These people were hunter-gatherers, although that label may conjure some inaccurate images of scattered bands of opportunists foraging about in the bush. More accurately, these were buffalo people who followed and worked the great herds. The hunting of their economy was not based on chance encounters with lone beasts, but was instead a studied relationship with the herds. The stone-tipped spears that did the killing were not so much weapons as butcher's tools, brought in to the slaughter once the herd had been corralled.

This early buffalo society built its economy before the days of the domesticated horse. Prehistoric horses were extinct and would not return until the Spanish came. These were the dog days, with the domesticated wolf in the role of beast of burden for the mobile lives of a people following the herds. Hunting relied more on wit than on the speed of the horse or the efficiency of a particular weapon. Bison were customarily dispatched in jumps or surrounds. Using a variety of decoys, natural barriers, fire, and a knowledge of the habits and usual movements of the herds, the hunters would maneuver a group of bison into a narrow box of a canyon to corral them or haze the herd over a cliff; the animals would in panic stampede to their deaths. The aptly named Head Smashed in Buffalo Jump in Alberta was used in the latter technique for more than seven thousand years. Such jumps were still in use at the time of white settlement, and an early account by a Hudson Bay trader living among the Blackfeet of the northern Rockies reported 250 bison taken in a single jump. Archaeologists working in Colorado have excavated an 8,500-year-old site containing the carcasses of 152 bison, wedged together in the heap of a single hunt.

Those animals not killed in the crush of the jumps were dispatched first with spears and later with the bows and arrows that worked their way into plains culture about 1,400 years ago. The spears were not simple devices; even those of the Clovis people featured removable hafts so that when an end broke off in an animal the hunter would simply reuse the long main shaft by attaching a new haft and point. Stone points often chipped or broke, but were simply resharpened to make a new and smaller point. The atlatal, a special sling for throwing a spear, was important to the continent's technology almost from the beginning. Modern archaeologists have duplicated the whole array of stone tools and weapons, and through trial and error have become adept in their use, even once butchering a zoo elephant that expired of natural causes.

The size of the kills suggests that the business of butchering consumed a great deal of effort. It must have been slow work whacking up a couple of hundred bison, but also something of a festival. These kills couldn't have happened very often, but must have flowed with annual cycles tuned to take advantage of the movements of the herds. Perhaps there were smaller, opportunistic kills during the rest of the year, perhaps of other animals such as elk and deer, but these great kills must have been the mainstay and defining event for the tribes, gathering the whole of the people together to work a year's worth of meat.

There is some mammoth-day evidence to suggest early hunters simply cached the meat in the fall near the frontier of the glaciers in pits that were early deep freezes. Later the people came to rely heavily on drying for storing the meat. In particular, the plains culture developed pemmican, which was a mixture of pounded flat and dried meat and melted fat. The pemmican was sewn into bags and would keep for months. Pemmican eventually became popular with European invaders. Such was the efficiency of this ancient traveling food that the English began importing it to use as fodder in the prosecution of continental wars. From the beginning, though, the plains people used pemmican as a trade good. Pemmican leaves no archaeological record, but those goods that do remain suggest there was a strong trading tradition throughout the continent. Remember that five-hundred-year-old mask made of Atlantic Ocean seashells that was found in Montana's Sweet Grass Hills.

There probably were variations in social patterns through the millennia, but on the whole the evidence suggests that plains culture was not a pure and isolated form. Rather, there was a continuum of ways of life that ebbed and flowed as grass culture flowed with drought cycles. There were cities such as those of the Hopewell culture and the later Mississippian people along the eastern edge of the plains, a settled agricultural world based on maize, beans, and squash. Archaeologists have calculated that about ten thousand people lived in one mound city near present-day Saint Louis. On the southwestern edge of the plains arose the cities of the cliff dwellers and pueblo dwellers, also based on farming.

The presence of both cities and bison kill sites in the region sets up the two poles of humanity that is not at all unique to the American plains, but a story as fundamental as that of Cain the cultivator and Abel the herder. Yet between these two poles there was no sharp border. Rather, the relationships with nature and food varied with the fortunes of the grass and the technology to exploit it.

In the driest places of the plains, in the places of pure grass and bison, the hunters prevailed and hunting was the dominant, if not the lone factor in an individual life. The Blackfeet exemplify the purest of this form, if only because their life was the most inscrutable to the agrarian, tree-loving European eyes that found them. The colonists held them as among the most savage of the savage, so attuned were their lives to an alien world. They were a meat people. Their word for meat translates as "real food," just as their word for themselves translates as "real people." They ate plants but called them "nothing food."

Imagine Europeans encountering plains people feasting at a fresh bison kill. While the meat eaters worked, they ate, pulling the choice parts off knives, as children elbowed in for the best. They ate the liver, kidneys, tongues, eyes, testicles, fresh fat, marrow from leg bones, and the hooves of unborn calves. Some tribes first removed the stomach from a freshly killed bison, filled it with blood, pieces of liver, and other choice bits, suspended it, and cooked the mass by adding hot rocks. The intestines were diced up and dipped in bile, then fed to children like candy.

These were practices that the early white explorers, using a label that revealed as much about the labeler as the labeled, called "barbarian." The whites were revolted, and resolved to survive the plains on a diet of pure muscle meat. The whites got sick and stayed that way until they took the advice of friendly Indians. For instance, once the whites learned to eat the tallow from around the kidneys, they found the persistent sores that had developed in their mouths immediately healed, just as the Indians said they would. A meat life requires adaptation.

By the time of white settlement, there was a core group of mostly nomadic plains tribes, the meat culture. These were the Arapaho, Assiniboin, Blackfeet, Cheyenne, Comanche, Crow, Gros Ventre, Kiowa, Kiowa-Apache, Sarsi, and Teton Sioux. They formed the heart of the buffalo culture and shared the unifying religious ritual of the Sun Dance. Then there were tribes considered seminomadic that hunted bison certain times of the year but also relied heavily on some agriculture. These were the Arikara, Hidatsa, Iowa, Kansa, Mandan, Missouri, Omaha, Osage, Oto, Pawnee, Ponca, Santee Sioux, Yankton Sioux, and Wichita. On the fringe of the plains, there was a group of tribes that hunted bison only occasionally and were ensconced in permanent settlements or hunted mostly in the wooded upper Midwest. They were the Plains Cree, Plains Ojibwa, Shoshones, Caddo, and Quapaw. This pattern did vary through the years, but the overarching rule was nomadism in the arid grasslands, with a tendency toward settlement and agriculture as one moved to the temperate edges, either the mountain valleys of the West or the tall grass, woodlands, and rain of the Mississippi valley.

Besides climate and the grass, the force that most significantly changed this shading of life ways from settlement to nomadism was technology, or more to the point, the arrival of the horse and its three-hundred-year spread through the plains. The return of this animal to its native land spawned a revolution in plains culture, as did the people who brought it. Wallace Stegner points out that plains Indian culture, at least as Europeans found it in the nineteenth century, was a "white creation." A variety of factors makes this true, but the central player is the horse. In the space of a couple of hundred years, the plains tribes took to the horse as if they had invented equitation. The days of the buffalo jump and surround faded, replaced by the chase. With horses, bison were killed singly, lessening the requirements for organization and social structure. The hunters no longer killed a cross section of the population, but selected young animals and their choice meat.

Wealth and prestige went along with ownership of horses. Horse theft and the military prowess it required became the system of redistributing wealth. Overall, nomadic life became more tenable, and so there was a shift in culture. Seminomadic people became fully nomadic, such were the attractions of the life. Tribes such as the Sioux that had been at least partially agricultural and settled abandoned their villages and took to the plains on ponies. In all of this, plains Indian culture came more and more to resemble grassland culture worldwide, and in doing so set itself up for a confrontation with the approaching Europeans, who were militantly agrarian. During this confrontation in the late nineteenth century, Cain's and Abel's ancient battle lines would reemerge.

Grassland culture is nomad culture owing to the simple fact of its aridity. Agrarian culture exists at its fringes or along river courses or under irrigation. Farmers generally served at the bottom of the pyramids that were the great autocratic societies. They were the many pillars necessary to support such social projects as irrigation and were necessary so that the elite might bleed off surplus grain as taxes to support palaces, art, writing, and commerce. We know the most about these settled societies simply because they left a written record. Writing was first necessary to record the transactions of grain, later necessary for priests and poets. Literature is a function of grain. Nomadism leaves no written record; it is illiterate. Instead of writing, nomads accumulated sheep, hides, camels, and horses, especially horses, and stories.

Nomadism implies motion, not a random motion of the nomads' choosing, a wanderlust, but a motion dictated by the conditions of the land. Nomadic economy is based on wild grass too thin to support pastoralism, the fixed pastures of domestic animals, or the commons that grew up in conjunction with settled agriculture. Pastoralism, the intensive raising of domestic stock, is an intermediate step between farming and nomadism. The nomadic economy is extensive rather than intensive. Nomads follow animals following grass.

The crown-of-creation fallacy from evolution works its way into cultural anthropology. The fallacy is based on the notion of progress and assumes the line begins with hunter-gatherer nomads and ends with cities. It suggests nomads were nomads simply because they lacked the technology or skills or intelligence to become farmers, and given the opportunity, would have done so. The record suggest otherwise: Nomadism is a condition dictated by the land.

The writer Bruce Chatwin, a lifelong walker whose peregrinations eventually killed him by way of a disease he caught walking in China, was a defender of nomads and suggests that through history there has been a dynamic equilibrium established between nomads and farmers, a tension, a cultural conundrum. This tension lies deep even in our culture and arises in one of our central myths, the fratricidal battle between farmer Cain and herder Abel, the first children of Adam and Eve.

The origin of this myth itself is a good example of the dynamic relationship that existed among wandering peoples. The story arose at the basis of the Judeo-Christian tradition among the Bedouin people of the Middle East during what is known as the patriarchal period. This was a time of substantial shifting pressures in the Middle East, similar to the arrival of the horse in the New World. Some of the herders about this time, just more than three thousand years ago, had managed the domestication of the camel, a revolution that weighted the world toward pure nomadism. Meanwhile, agriculture had begun to build great cities.

The Hebrews arose in the space between and felt the tug of both city and nomad. They formed a society the German anthropologist M. B. Rowton has labeled "dimorphic," neither city nor nomadic, but a seminomadic collection of herders, most of them tied to specific villages. This society internalized the great dichotomy that would reverberate across the globe on into this century, when on this continent Cain would again kill Abel.

Writes Chatwin:

> Nomadism is born of wide expanses, ground too barren for the farmer to cultivate economically—savannah, steppe, desert and tundra, all of which will support an animal population providing that it moves. For the nomad, movement is morality. Without movement, his animals would die. But the planter is chained to his field; if he leaves, his plants wither.

The conditions that dictate the balance between either form of life periodically shift in periods of drought or excess or social upheaval. During these times, the line between cultures shifts. Wanderers leave the overpopulated cities. Nomads beset by drought trade or steal grain surpluses from farmers as protein for their herds. Irrigation fails and sends the city dweller to herding. Even during normal times, active trading networks form between the two sides. Nomads need the city's tools and grain. Cities need meat and hides and markets. Writes Chatwin, "A nomad independent of settled agriculture probably has never existed." This economic interdependence, the fact that each pattern of living exists as a sort of safety valve for the other, is lost in the literature. Civilization reserves for the nomads its epithet "barbarian," a word that serves civilization in its periodic raids and wars on grassland.

This tension between civilization and nomads erupts wherever grasslands exist. The Masai of Africa, herdsmen with the habit of rounding out their diet by bleeding their cattle, still suffer it. The Hebrews' way of life earned them enslavement by the settled irrigators, the Egyptians. From the Egyptian point of view, they may have earned the chains. "Raids are our agriculture," says the Bedouin proverb.

It seems, though, that the most protracted war between agrarian and nomadic groups played out on the Asiatic steppes, both in the west, where Mediterranean cultures in Greece and Rome confronted steppe people along the Danube, and in the east, where the most entrenched of civilizations, the Chinese, confronted the same people, the barbarian Mongolian hordes. The Mongolians even appear in Christian tradition as the horsemen of the Apocalypse, the kingdoms of Gog and Magog.

In A.D. 376 the mounted archers of the grassland steppes, the Huns, were sacking the Gothic kingdoms of eastern Europe and sending their refugees to collapse the Roman empire. A contemporary Roman historian, Ammianus Marcellinus, wrote of the Hun: "No one ever plows a field in their country or touches a plough handle. They are ignorant of home law, or settled existence, and they keep roaming from places in their wagons."

The Chinese said of the same people, "In their breast beat the hearts of beasts . . . from the most ancient times, they have never been regarded as a part of humanity." The Chinese had been dealing with the Mongols for centuries and had built the Great Wall to suppress periodic barbarian raids, or as some suggest, to keep their own subjects at home. The farmers bent under the autocratic rules of irrigated society sometimes fled their plows and adopted the roaming life of the Hun, just as Indians in the Mississippi valley did when horses arrived.

In our imagination, we see the Great Wall as the firm line between an agricultural, literate society and barbarian nomads, but the American Owen Lattimore, who spent the early part of the twentieth century living and traveling among the Mongolians and Manchurians of the steppe, said the line was never so firm. In fact, in Mongolia, far beyond the Great Wall, one still finds traces of long-buried walls built perhaps by the Chinese, perhaps by the Mongols themselves. Lattimore says these walls likely marked the advance and retreat of agriculture through the centuries as farmers attempted to take the grasslands and failed.

An American soil scientist produced one of the best explanations of these northern walls. They correspond almost exactly with a shift in soil type that made agriculture possible on one side, tenuous on the other, during certain climatic conditions. Perhaps even the Mongols themselves did some of the farming, as various leaders saw the wisdom of diversification. Through history, Mongolia has never been simply a sea of horsemen, camels, grass, and sheep. When the Italian explorer Marco Polo ventured through the region in the thirteenth century, he found cities, traders, craftspeople, princes, Moslems, Christians, and Buddhists. We can only imagine the cities, but perhaps they were much like those we would have found in the American plains at about the same time, the trading centers of a grassland economy.

Certainly, the dynamic equilibrium between agriculture and nomadism always played out in microcosm throughout the variations in conditions of the steppe. This tension between Mongols and Chinese can be read as the advance and retreat of warring people or the advance and retreat of grass. Lattimore tells this story of finding some abandoned grain-working tools alien to Mongol culture:

> Once when I was traveling by cart with a Chinese-speaking Chahar Mongol along the edge of cultivation we rattled into a Chinese village where we saw an old stone roller, reclaimed by the [Chinese] colonists and brought into use again. "The Chinese tell us Mongols," my companion said, "that when they colonize our land they are not really taking what belongs to us but taking

back what used to belong to them. Wherever a millstone or mortar or stone roller can be found, they say, the Chinese will one day come back."

What mediated the balance between Mongol and farmer was the horse. As the embodiment of the motion of the grassland the horse was in real ways the embodiment of the power of the land. Grassland people are horse people.

The first man to sit a horse probably lived about six thousand years ago on the western edge of the Asiatic steppes in what is now Ukraine. In that same region and among those same people there arose a horse-based technology, first with invention of the wheel and cart and later, about four thousand years ago, the invention of the spoked wheel that made the chariot possible. This leap in technology would later make these steppe peoples the scourge of the civilized world.

A second equestrian tradition eventually arose among the Old World's other most prominent group of nomads, the Arabs. When Coronado first saw Indians, he compared them to Arabs.

The Arabs, whose very name meant "tent dweller," developed one of two dominant strains of horseflesh in the Old World. The more common strain, the original northern strain of the steppes, was called "cold-blooded." Through most of their history of stalemate, the Chinese had contended with Huns on cold-blooded mounts. But during the reign of the Emperor Wu-ti, 145–87 B.C., certain Huns got their hands on hot-blooded horses, horses so holy they were rumored to sweat blood, and came to be known as "heavenly horses." So important was this change that Wu-ti would devote a lifetime of warfare, treachery, bribery, and cunning trying to get his hands on heavenly horses. They were the grassland equivalent of thermonuclear weapons, especially in the hands of the Huns.

"The Huns, we are told, bought, sold, slept, ate, drank, gave judgment, even defecated without dismounting," writes Chatwin. This attachment to horses is the quintessential regard for animals that mediate the nomads' relationship to the land. Because of the dictates of travel and the nomad's needs for ever more children, youngsters were not nursed by their mothers long but were fed mare's milk and sat on a horse literally before they could walk. The attachment to animals often equaled or even superseded that to human life.

This Mongol way of life held through all the millennia until the railroads came and upset the balance between grass and grain. A railroad is industrialized motion. Suddenly markets shifted and the trade network became global. Capital began supporting farming ventures deep in Mongolia. Supplementing the rails of steel was more steel, the rifles and trucks of both the industrialized Chinese and Japanese. Lattimore watched all this happen, noting that the end of steppe life came with the end of "the superiority of the nomad in speed and mobility." Mobility now was in the power of the Chinese:

> Modern arms gave the Chinese an immediate military advantage, but railways gave them a permanent advantage. By making possible the export of agricultural produce over distances which had been prohibitive in the age of carts and caravans, they enabled the Chinese to settle permanently in Mongol territory.

Tom Lawrence, an American plant breeder and a man of the plains who ought to have understood the irony of his statement, but apparently did not, said this:

> Traveling in Inner Mongolia in 1982, I saw communes where the modern Mongols live in mud houses surrounded by their camels, donkeys, sheep and cattle running all over the place. Of course they used to be nomadic, lived in portable yurts and had a traditional very effective grazing pattern that kept the range in shape. But when the communists took over in 1949, they said, "Comrades! You cannot travel around; you gotta stay in place!" So the Mongols settled down. They cut down all the trees to

make their mud house rafters; they held every animal they owned on communal pastures, and pretty soon they had a desert. Now they're sowing North American cultivars of their native grasses.

Early in the nineteenth century, the artist George Catlin lived among the still nomadic tribes of the American plains. He wrote, "They live in a country wellstocked with buffaloes and wild horses, which furnish them an excellent and easy living . . . and they are among the most independent and happiest races of Indians I have met with."

The railroads ended this life. Like the great grasslands elsewhere, the American plain was to see a replay of the battle between freedom and the plow.

Several die-hard myths of the American West drop dead at the sound of the single French word *Chadron*. It actually is not the original French but the meaning is there, with the etymology of a word that is like an archaeology of a place. The myths it dispels are these: First, that the homesteading of the grasslands was the beginning of their economic exploitation and the establishment of an economy of the West. Second, that Europeans could not understand or integrate their lives with the lives of the plains people that came before them. As far as Europeans go, the French settled the northern tier of the West; the subsequent waves of English, Scandinavians, and Eastern Europeans unsettled it.

Chadron is a town in the northwest corner of Nebraska, just south of the Black Hills, where they dug up the gold, just south of Pine Ridge, near the place where the Sioux leaders Crazy Horse, Big Foot, and Sitting Bull all died, all murdered after the surrender of their people. In 1882, just to the north in South Dakota's badlands, Sitting Bull himself took part in one last buffalo hunt just after the great herds disappeared.

Chadron now is a farm town, a ranch town. A small college stands at its hub. Likely few of its residents, descendants of the few homesteaders who managed to hang on, can find resonance in the town's odd French name. Most are Scandinavian or Eastern European, descendants of a later white wave. To hear much French, a person needs to go north to the Pine Ridge reservation or the adjacent Rosebud reservation, where phone books read at least halfway like phonebooks in New Orleans or Montreal. Aimiotte and Pourier mix liberally among family names like Red Cloud and Her Many Horses. The Sioux derive their French surnames from the same source as Chadron, from French trappers and fur traders.

Chadron is a particularly apt place to work this vein of western history, largely owing to the presence of Charles E. Hanson, a one-legged longtime Republican and refugee from the Nixon administration who has spent a second life believing his Swedish surname does not tell all there is to tell about his native Nebraska. Hanson's an old man now, the leg lost to a disease of age. Most of his earlier working life was spent administering the Department of Agriculture's public works program in Washington, D.C. He served under Lyndon Johnson, which "cured me of being a Democrat," but quit the business altogether in 1969, when he found he could not stomach the Nixon administration. He still speaks bitterly of politics.

Hanson came home to Nebraska to be a historian and now is the director of the Museum of the Fur Trade that sits in a stockade-like masonry fortress just outside of Chadron. The building houses not a tourist trap but a world-class collection of artifacts from the fur trade. The collection is not simply local; rather it traces the four centuries of trapping and trade from the original East Coast beach head to Russian commerce with Eskimos and Aleuts. The building holds Sharps rifles, colonial hats, and translucent Eskimo raincoats made of seal intestine.

The bulk of this trade was handled by the French, who both acted in national interests and later were hired as individuals by British and American fur companies. The French made their original inroads to the West, both economically and genetically, through the upper Great Lakes region and on into Minnesota, where the Sioux lived then. They trapped and traded for the British through the Hudson's Bay Company and for the Spanish to the south. Seventy years before Lewis and Clark, the French already were common all along the Missouri basin, they themselves preceded by the Spanish. French traders visiting a Mandan village on the northern plains found Spanish-speaking Indians there in 1742. Jefferson cemented his control over the Louisiana Purchase in part by using the Choteau family, powerful traders who started Saint Louis, near the site that had been used by earlier Indian cultures as the center of a trade network. "The Choteaus in Saint Louis were a pretty sharp bunch of frogs, I'll tell you that," says Hanson.

Chadron gets its name from a trader, probably named Shattron, who set up a trading post there in 1833. The post fit into the trade network of the mountain men working the beaver pelts on into the Rockies, but was situated to trade in bison hides and robes. Robes were just another trade good in 1833, but within forty years the trade would reach a fever pitch and remake the West.

The point to be made about the early fur trade, at least in bison robes, is that it was stable, both in the economic sense and in the environmental sense. It was appropriate technology. The network of trade had existed over most of the continent at least since the beginnings of white settlement. The prime motive for French, English, and to a lesser degree, Spanish and Russian exploration, it spawned the legendary monopolies, including the imperial Hudson's Bay Company and Astor's American Fur Company. So deeply is leather rooted in our economy that we call a dollar a buck today only because that was the standard price for a deerskin in the American South.

The native hunters of the continent were always heavily integrated in the flow of goods. Hanson doesn't believe the nomadic tribes of the grasslands had much to do with the business when the object was beavers, largely because they would have been loath to wade around in half-frozen water setting traps. Around the turn of the nineteenth century, however, the French traders began dealing more earnestly in the softened buffalo robes the nomadic tribes had used for centuries. These were not tanned hides, because the tribes had no tannic acid, but rather hides treated with the brains of the animal, then pounded and softened like buckskins. The robes likely had been an important trade good even before white incursions. A firmly established pre-Columbian trading post near Horse Creek in eastern Wyoming featured white fox skins, carved walrus ivory, seashells, pipestones from the Midwest, and products from Mexico.

The lure for business with the whites later was largely based on mobility. Hanson said the plains tribes eagerly swapped the buffalo robes, which weighed about ten pounds, for woolen Hudson Bay blankets, which weighed about four, for about the same reason a backpacker eagerly trades a flannel sleeping bag for a lighter one of goose down. Lead, gunpowder, iron knives, and tanned leather also figured heavily in the dealing.

This new element of trade with the whites produced only a minor shift in plains life. Hanson says the traders then preferred hides from cows while the tribes, now used to the selective hunting that the horses allowed, preferred the meat of young bulls. Supply, however, adapted to demand, and the Indians began taking cows, but no more animals than before the trade.

A family would sell ten to fifteen robes a year, says Hanson, depending on how many wives there were in the family. Labor, more than demand or the Indians' ability to kill bison, was the limiting factor of the trade. A family could sell only as many hides as it could process, and the sale was incidental to the meat the family used. The whites did have the labor to soften or even tan buffalo hides in eastern cities, but green hides were far too heavy to transport across the plains.

This trade system gradually worked its way into both white and Indian economies. By 1827 buffalo robes were selling briskly in eastern markets, advertised as carriage robes. The trade created a stable hybrid culture, a model of successful integration of Europeans into the nomadic culture of the plains, a people we call *metis.* This French word encodes exactly the same notion as the Spanish *mestizo,* an intermarrying of Indian and European, both culturally as well as genetically. The *metis* were, in the vernacular of the day, half-breeds, the union of the French fur traders and the tribes of the Great Lakes region.

Some of these people made their way west in distinctive squeaking wooden-wheeled wagons called Red River carts. They founded villages in what is now southern Alberta and northern Montana, similar to the seminomadic villages of plains tribes and even ancient Hebrews. Presaging arguments on patterns of western land use that would arise and build toward the end of the nineteenth century, arguments we will examine in detail later, the *metis* organized their villages along rivers to allow a limited agriculture in the arid land, small garden plots in the valleys. The bulk of the surrounding land, however, was held as a commons. Here the squeaking carts and their people ranged as nomadic hunters, taking their own meat and buffalo robes.

By 1820, professional white hunters moving into the upper Missouri stretches of the Montana plains found the Red River carts already in control. Their annual expeditions out of Alberta brought about 610 carts and took 146,000 buffalo a year in the period 1820–25. By 1835–40, the expeditions averaged 1,090 carts and 212,500 buffalo a year, with the robes shipped southeast through Saint Louis.

Aside from the *metis* hunts, the Indians were killing about 3.5 million bison a year in the decade before the Civil War. Estimates place the total bison population of the plains at the time between thirty and seventy million, with the most detailed estimate standing at fifty million. The kill and the trade were sustainable in any sense of the word. Just a decade later, however, that would change and change quickly. The bison hordes of the plains would become virtually extinct in a space of a half a human lifetime. Ironically, what killed all the bison in this grassland, ruled as it is by freedom and mobility, was motion, locomotion, locomotives.

We cannot weave for ourselves a fully satisfying answer as to why the Indians did not wipe out the bison, even given access to American and European markets. A fully satisfying answer would have to account for motive and ethic wrapped deeply in a people who are willing to drag bison skulls attached to slits in their muscles by leather thongs. An answer would be somehow wrapped in the Sun Dance, and the understanding of this is lost to literate and literal Western minds. All we need to know is as ephemeral as the running wind and is not, in our civilized sense of knowing, knowable.

Yet we can know enough to cement a partial answer. The native nomadic people did have the technology and knowledge to exterminate the bison; they used the same rifles that the white hunters fired into the herds until their barrels warped with the heat. On occasion, the Indians were completely capable of hunting wantonly, especially after the whites discovered how well whiskey lubricated the trade.

Much is made of plains tribe hunting taboos and their effects against overkill. The writer Tom McHugh describes the fate of a hunter who violated those rules: "After flogging him, marshals confiscated or killed his dogs and horses, cut his lodge into pieces, burned his tipi poles, broke his bow or gun, took his meat supplies, and ripped his hides, reducing him to total beggary." Examining these rules, however, indicates many of them were not so much designed to protect the bison as they were to ensure that independent operators didn't spook the herd and spoil the hunt. The buffalo culture depended upon cooperation and the rules were meant to ensure it.

Probably, we can leave rules, motive, ethic, and religion out of this argument. The bison and the plains tribes coexisted through the millennia simply because killing all the bison would have been an enormous effort and there was no reason to do so. Hunting was work, an expenditure of energy. The return on that expense of energy was food, clothing, and trade goods. A community could use only so much food and beyond that there was no point in finding, herding, killing, butchering, and storing the meat. As for trade goods, the medium of exchange was robes, and processing them was labor-intensive, again limiting the kill to available hands. There were likely a hundred buffalo to every plains Indian. How big of a dent could the Indians make?

We may believe that the plains people were ethically incapable of hunting the bison to extinction and that may be so, but beyond that, there was simply no point in doing so. Nor was there a point in the whites' doing so, at least not until after the Civil War, when revolution unsettled the plains.

The buffalo literally fell to the wheel. From lap robes for New Englanders, the uses of bison hides spread in America and Europe. Manufacturers began using some green hides for leather, which gained quick acceptance in vast markets. The British army favored the tough leather and went soled and saddled into Crimea with American bison. It became fashionable to panel the ostentatious houses of Europe's and America's burgeoning middle classes with bison leather, which also made fine book bindings and buggy tops. The real explosion of demand, however, stemmed from a certain elastic quality of the leather. This made it ideal for industrial belting, the stuff that drove the wheels, pulleys, and shafts that were remaking the Western world. The industrial revolution meant we began to see the natural world as a machine, and in this light, we saw nothing wrong with making animals into machines.

Buffalo leather required industrial tanning mills fed by green hides, not rawhide robes. Green hides weighed maybe fifty pounds apiece, five times as much as a robe. The Indian trade freighted on the backs of horses and the *metis* trade in Red River carts could not handle the load, so the slaughter of the buffalo was a simple matter of following the industrialization of the plains, following the railroad.

During the 1860s, President Lincoln both signed the Homestead Act and granted more than ninety million acres of western lands to the railroads as an inducement for them to push into the West. By 1867, market hunting of the bison was taking a quantum leap, worthy of the land-grab age. The extermination proceeded in three distinct phases, exactly coincident with the extension of the railroads into three broad areas of the country.

It began in the middle, in Nebraska and Kansas, with the great

Republican River herd. In 1870, two million bison were taken from the region. In response to this news, three legislatures—in Montana, Idaho, and Wyoming—passed bills outlawing the hide hunting, and a similar bill cropped up in Congress but died there. Fearing a loss of market, the railroads straightaway began concealing the extent of the trade.

By 1872, hide hunting was the largest industry of Kansas. Most of the trade was running through Saint Louis, but the trading post at Fort Dodge—later Dodge City—was on its way to becoming the hub of the slaughter. The writer Mari Sandoz offers an account of an incursion into the region made by the place's namesake:

> In May 1871, Col. R. I. Dodge drove in a light wagon from old Ft. Zarah to Ft. Larned on the Arkansas. At least twenty-five of the thirty-four miles were through one immense dark blanket of buffaloes—countless smaller bunches come together for their journey north. From the top of Pawnee Rock, Dodge could see six to ten miles in most directions, all one solid mass of moving animals. Others who saw the herd reported that it was twenty-five miles wide and took five days to pass a given point—probably fifty miles deep. Dodge estimated that there were about four hundred eighty thousand in the one herd; perhaps a half million that he saw himself on that single day. With those that others observed beyond Dodge's sight but still of the same herd, it was estimated at from four million to twelve million counting fifteen head per acre for the former number. This was the great southern herd.

By New Year's Day of 1873, the Santa Fe Railroad completed a phase of work near Dodge City and laid off its crews, creating a glut of men needing money. By spring there were at least two thousand hide hunters in the region. One party of sixteen men killed 28,000 buffalo. The railroads admitted to shipping 1,378,359 hides between 1872 and 1894, but General Nelson A. Miles reported at the time that the kill was closer to 4.5 million.

The great herds of the middle region were wiped out then, and the kill moved south to Texas for its second major phase. Fort Worth replaced Dodge as the trade center only three years after trade at Dodge had begun. In the fall of 1875, Tom Nixon set a new record for the southern hunters. From September 15 to October 20, he alone killed 2,173 bison. The following year a London newspaper would label the hunting "a scandal to civilization," a protest that was echoed on this continent. But it was already becoming clear the scandal ran much deeper than an assault on wildlife. It was not so much a scandal to civilization as it was the imposition of civilization.

Protests of the destruction of the southern herd issued not just from London but from Texas, where the state legislature appeared ready to pass a bill outlawing hide hunting. Then General Phil Sheridan, a leading figure in the Grant administration's war on the plains Indians, lobbied for the hide hunters in a statement that was a manifesto of civilization's agenda. He told the legislature:

> These men [hide hunters] have done more in the last two years and will do more in the next year to settle the vexed Indian question than the entire regular army has done in the last thirty years. They are destroying the Indians' commissary; and it is a well-known fact that an army losing its base of supplies is placed at a great disadvantage. Send them powder and lead, if you will; but for the sake of lasting peace, let them kill, skin and sell until the buffaloes are exterminated. Then your prairies can be covered with speckled cattle, and the festive cowboy, who follows the hunter as a second forerunner of an advanced civilization.

It is easy to make too much of a statement such as this, as if the extermination of the bison were the product of a willed agenda. The agenda was superfluous. The technology and economic core of the matter were enough to bring about the result: vast markets, a vast and relatively unexploited resource, and finally the railroads to connect them. The laws against hide hunting were as unenforceable in nineteenth-century America as laws against ivory poaching are in twentieth-century Africa. Industrialization drives extermination.

By 1877 the Texas bison had retreated to the Staked Plain, the Llano Estacado of Texas and Oklahoma. Among the southern plains tribes, it is widely believed that the bison first sprang from the earth in the canyons here. Archaeological evidence indicates that both bison and bison hunting are ancient in Texas. In 1877, the place was deep in one of the West's periodic droughts.

Then there came a meteorological oddity, a waterspout that poured in across the land and built a great shallow lake. The thousands of bison of the Texas herd had been well away from the lake when it formed, but evolved as they had been in a place of drought and wind, they caught the scent of water on the air and began a great motion, until the broad shallow lake became a wallow of virtually all the southern herd.

The hide men followed a few days later, at first thirsty themselves, hectoring off bison and wolves so that they and their horses could drink. Then they killed. A pair of hunters took 6,200 hides in the next few months. A lone man working to the south of them took 4,900. All around the Llano it was the same; wherever a seep or draw popped up, the bison congregated in their evolved defense against drought, a defense that rendered them helpless against a new enemy. The new enemy was not man, taking drought-bunched animals here just as he had in Clovis times when the mammoth bunched around that water hole in Arizona's San Pedro Valley. The enemy was this new man, whose railroads, like a lens, could focus the hungers of men of an entire world to burn on a single spot in Texas.

By 1878, the southern herd was gone, save a relict band that was to roam the plains until a cowboy spotted and fired on them in 1886. A hunting party took all of them, fifty-two animals, that winter.

• •

In 1876, the Northern Pacific Railroad was ready to punch into the Yellowstone country of Montana, but was stalled at Bismarck, North Dakota. Much of what lay beyond was legally Indian lands. Buffalo trade had existed there since at least 1820, but under Sioux, Blackfeet, Cheyenne, Crow, and *metis* rules. This native tenancy disallowed settlers and railroads.

George Custer, however, had by then died on the Little Bighorn, a tributary of the upper Yellowstone, and inflamed opinion gave the railroads the boost they needed. With the Sioux leaders off in the aftermath of the battle, there remained only a few warriors in camps to conduct business. These the army arrested and jailed; they were told that the women and children left outside would not be fed until they signed a treaty relinquishing their claims to the upper Yellowstone and Powder rivers.

By 1881, the railroad had pushed upriver to the area of the herds. The same year, 75,000 hides were shipped on the Northern Pacific. By 1882, there were five thousand hunters on the northern range. By 1885, no more hides were being shipped out of the Yellowstone country, habitat of the continent's last great herd. The industrial hide hunting that had begun eighteen years before had ended. There were, for all practical purposes, no more animals to kill.

The Missouri joins with the Yellowstone at the far eastern edge of Montana in a landscape forlorn and vacant by standards of anywhere else but the West. Nearby, the government is restoring Fort Union, an early outpost of civilization. That elk we first encountered leaving Montana's Sweet Grass Hills would have passed nearby, if it did indeed follow the Missouri all the way. The glaciers of the Ice Age came to its edge, then stopped, as if marking the high plains' vital line. It seems as if a good bit of the place's history, from the glaciers to the single elk, have honored the line.

In 1886, William Hornaday, chief taxidermist for the U.S. National Museum, heard that the bison were imperiled and became alarmed, largely because his museum had not collected suitable specimens. He came to Miles City and there hunted the land to the north that is the hydrological divide between the Yellowstone and Missouri. With hired guides Hornaday hunted eighteen days before spotting bison, a relict herd that the museum keeper finally killed in the interests of science. Unable to finish skinning part of their kill, the party left it overnight and returned the next morning. Hornaday wrote:

> When we reached it we found that during the night a gang of Indians had robbed us of our hard-earned spoil. They had stolen the skin and the eatable meat, broken up the leg bones to get at the meat and even cut out the tongue. And to injury the skulking thieves had added insult. Through laziness they had left the head unskinned, but on the one side of it they had smeared the hair with red war-paint, and the other side they had daubed with yellow and around the base of one horn they had tied a strip of red flannel as a signal of defiance.

Hornaday's specimens were returned and exhibited back East with the come-on of "real buffalo grass, real Montana dirt and real buffaloes." In 1887, the American Museum of Natural History decided it too would like suitable specimens before extinction, but a properly equipped hunting party detailed to the same region found nothing.

Only a human lifetime before, in 1819, the same region of the upper Missouri was the destination of steamboats—loaded with explorers, traders, and scientists—fitted out under General Henry W. Atkinson at Saint Louis. Wrote an observer as the expedition departed:

> See those vessels, with the agency of steam advancing against the powerful currents of the Mississippi and Missouri! Their course is marked by volumes of smoke and fire, which the civilized man observes with admiration, and the savage with astonishment. Botanists, mineralogists, chemists, artisans, cultivators, scholars, soldiers; the love of peace, the capacity for war: philosophical apparatus and military supplies; telescopes and cannon, garden seeds and gunpowder; the arts of civil life and the force to defend them—all are seen aboard. The banner of freedom which waves over the whole proclaims the character and protective power of the United States.

With the fire and the steam that enforce the new motion, it is all here in this passage, all that would bring the end of the bison and the people for whom the bison was food, shelter, and spirit. But as we follow this trail, we must direct our attention to the quieter forces in the weave, to the surveyors' telescopes, to the cultivators, and especially to the seeds.

5 Gridlock

It is the lot of the poet and maybe the politician to suffer the shortcomings of an earlier generation's science. Thought accretes like limestone from a sea, the geology of our ideas, and becomes the blocks we quarry to rebuild ourselves. The ideas that built America's nineteenth century accreted from rationalism but metamorphosed to yield romanticism and the progressives, and all these combined to build a structure, literally visible, on the landscape of today.

By the late nineteenth century we can hear the poets' part of this process and mark its larger path in their songs, especially in the voice of Walt Whitman, the chanticleer of democracy. *Specimen Days* was a prose account in 1881 of the poet's first and only journey into grassland, well after he first published *Leaves of Grass*. In it he says:

> Grand as is the thought that doubtless the child is already born who will see a hundred millions of people, the most prosperous and advanc'd of the world, inhabiting these Prairies, the great

Plains, the valley of the Mississippi, I could not help thinking it would be grander still to see all those inimitable American areas fused in the alembic of a perfect poem or other esthetic work, entirely western, fresh and limitless—altogether our own, without a trace or taste of Europe's soil, reminiscence, technical letter or spirit.

By grass, Whitman meant democracy, gentle and uniform fingers of human souls to spread across the earth like lawn. The real prairie America found and took was nothing so much as a vast blank sheet on which the nation would fill in the ciphers and faultless geometric proofs, later the poems, that would validate our experiment. At first these writings were simple, straight lines drawn across the earth, lines as naive, as confident, and as true as the geometry that inspired them. Rationalism would refine them. The lines would become logic's prayer to the prairie gods. On this land, America would eventually try to write a binding contract with creation.

Whitman sings of this in "Starting from Paumanok":

Victory, union, faith, identity, time,
The indissoluble compacts, riches, mystery,
Eternal progress, the kosmos, and the modern reports.

This then is life,
Here is what has come to the surface after so many throes and convulsions.

How curious! how real!
Underfoot the divine soil, overhead the sun.

See, the revolving globe,
The ancestor-continents away group'd together,
The present and future continents north and south, with the isthmus between.

See, vast trackless spaces,
As in a dream they change, they swiftly fill,
Countless masses debouch upon them,
They are now cover'd with the foremost people, arts,
institutions, known. . . .

Americanos! conquerors! marches humanitarian!
Foremost! century marches! Libertad! masses!
For you a programme of chants.

Chants of the prairies.

One can assume that William Quayle, a writer altogether as eccentric as Whitman, must also have had the Good Gray Poet as well as Whittier in mind when a generation later he wrote of poets in the plural:

> Once when Whittier mentioned the prairie grass he mismentioned it. If these poets had a word to say, it was in privacy. They spoke of this chiefest beauty of our continent as chancing to think of it while they were discoursing of something besides. What could have ailed them? For one thing, they were mainly seaboard poets. They knew the hills, the streams, the mountains, the sands and marshes of the sea; but prairies were not among their fellowships. They staid too near at home. They did not journey to the West far enough, or else they did not stay long enough to get the prairie wonder in their blood.

This is not a literary debate. Perhaps none of the combatant literati would have said it just so, but the bone of contention is this: Our science, our poetry, and our democracy fail because they lack specific information of the plants, but a notion as odd as this takes some telling.

Whitman's and the rest of the romantics' information did not arise from the land, but from the vapors of the European mind, and as a result, paradoxically, can best be read on the land today. There is ambiguity in the sentence "Whitman wrote on the land," but there is truth in the seemingly ridiculous literal meaning of it. This meaning may be read far more clearly in its original, not as it reverberates through poetry, but in the philosophical stones that built the poetry.

Thomas Jefferson, for instance. Jefferson did not invent the whole of the American story but in many ways simply repeated tales that had been told to him. Still he seemed to pull together a broader and more internally consistent layer of those tales than any other of the founders. From his mind devolved much of our notion of who we are, and much of his notion is so intimately tied to the land that it must be written on it.

If one were to read a synopsis of the American West one could do worse than to travel to Poplar Forest in Virginia, an estate that was Jefferson's retirement home and a reflection of the lines of his thought. It is a monument to geometry, with houses and gardens laid out as cleanly as on a draftsman's table. Geometry stood at the center of Jefferson's notion of harmony. The gardens at Poplar Forest are a reorganization of life to conform to geometry's lines.

And one could read a prologue of the West in Jefferson's garden at Monticello. The historian Donald Jackson reports that "Jefferson's fields and gardens contained hop clover, hemp, bent grass and winter vetch from England; alfalfa from the Mediterranean; Guinea corn and sesame from Africa; Nanking cotton from Asia; field peas, sainfroin, and turnips from Europe; sulla grass from Malta. His buffalo or Kentucky clover came from the inland grasslands of the Ohio Valley. All his prized orchard fruits—apples, pears, cherries, peaches—were introduced from abroad."

Jefferson never saw the grassland West, and it was not until very near his death that he ventured west far enough even to see the Alleghenies. He is, nonetheless, the founder of the West, if only by virtue of the Louisiana Purchase. The purchase was not, however, an isolated act. It was more a necessary factor in an equation that the founders believed would yield democracy as its product. It was a block in the nation's and Jefferson's stone wall of thought.

Jefferson was the chief carrier of what was in his time a wholly revolutionary idea, one we still cherish and have come to call the agrarian myth. By this we mean something more than that farmers are good. We mean that a particular type of farmer, the yeoman, who holds a small amount of land, sufficient for the care of his own family, but who neither works for wages nor hires others is a necessary precondition of democracy. This notion is what Henry Nash Smith calls the "master symbol" of the nation.

It is testimony to our continuing consensus on agrarianism that such people as actress Jane Fonda and singer Willie Nelson can rally in support of the same farmers championed by Senators Orin Hatch and Robert Dole. The myth is still the foundation of the agenda of the radical counterculture *Mother Earth News* back-to-the-landers and of the rock-ribbed Grand Old Party that holds forth from every feed store and Rotary Club of the Midwest.

The notion found its early voice in J. Hector St. John's *Letters from an American Farmer*, written in 1782: "We are a race of cultivators; our cultivation is unrestrained; and therefore everything is prosperous and flourishing." But more important, the land these cultivators owned would make them independent and therefore immune to the tyranny that had plagued Europe. The yeoman would be instead united by "silken bands of government."

Jefferson agreed, generally, thoroughly, as in a 1785 letter to John Jay:

> Cultivators of the earth are the most valuable citizens. They are the most vigorous, the most independent, the most virtuous & they are tied to their country & wedded to liberty & interests by the most lasting bonds.

Politics is power, and agrarianism is an attempt to invoke the power of the land, translate it as property, and so divide and confer that power equally among the many yeomen. The implications of this philosophy would firmly focus Jefferson's attention on the

American West for reasons he makes clear in a 1787 letter to James Madison:

> I think our governments will remain virtuous for many centuries; as long as they are chiefly agricultural; and this will be as long as there shall be vacant lands in any part of America.

Jefferson's theory of agrarian democracy was inherently expansionist and derived from a devotion to law. In his view, English common law was a direct result of Saxon incursions into early Britain, the natural product of free men pursuing free land. He believed that American colonialism was simply a continuation of the westward march of that process. The blessings of the West move west. Advances in democratic institutions would derive only from advances on the land.

Of course the lands he had in mind were vacant only in the colonizers' imaginations. The West had been successfully inhabited for ten thousand years, for several hundred by some Europeans. The lands were not even empty of agriculture. Jefferson was simply seeing the world through a particular filter of his time, a filter that survives and functions even in ours.

The rationalism that so profoundly influenced the founders, especially Jefferson, was a curious sort of hubris, mathematical and insistent on accountability. It was faith in reason above all else, even above observation. Given the verities as revealed by geometry and given the laws of reason, one could sit at a drawing table in one's library and deduce the nature of the world. In this, there was an article of faith that said nature eventually would add up, would reveal itself on all fronts as it had so cleanly in the geometric proof. There was an assumption that nature, like refined men, was rational, as if the calculations were the equivalent of a sort of contract. Once revealed through deduction, it became nature's contractual obligation to match the calculations: "indissoluble compacts."

A prime motivation of the Lewis and Clark expedition, for instance, was not just having a look at the newly purchased land. Jefferson settled on his explorers of choice only after the French botanist Michaux failed in a similar mission in 1793. (The mission became mired in some rather nasty espionage by the French.) Jefferson wanted a biologist, not a soldier, to make the trip because the expedition was to answer a biological question. Farmers in the East had unearthed the remains of animals no longer extant, especially mammoths, causing Jefferson to consider but reject the idea of extinction. A rational creator would not imperil order by bringing forth species simply to let them go. Jefferson's notion was to search the West for the mammoths that had for some reason left the East. Their discovery there would erase the illogical blot on nature's record.

More than biology, though, the rationalists' queen of sciences, geometry, opened the West. Geometry was necessary to draw the contract with nature, the invisible lines to government's silken bands. Jefferson's democratic equation was wholly dependent on the ownership of land. Not too much land, which would make for squires and tyrants. Not too little, which made for peasants and paupers. There needed to be a rational way to dice up the "vacant lands" west of the Mississippi, so there arose what we call the Jeffersonian grid, or rectilinear cadastral survey, as it was laid out in the Land Ordinance of 1785.

This took the navigators' lines of latitude and longitude and subdivided them in squares down to the township level, a square area of six by six miles. The thirty-six square miles were divided into sections of land of 640 acres each. These were quartered to 160 acres, called a quarter section, then requartered to 40-acre plots. These measures became the basis for parceling lands under the various homesteading and land entry laws. From the first, it was believed that 160 acres was exactly the amount of land a yeoman would need, and the later land entry laws of the 1860s all clung to this notion.

The survey drew the lines, and yeomen were to be dropped between the lines like numbers on a spreadsheet. The rows and columns were to sum as democracy.

Significantly, the whole scheme assumed a uniformity of nature in harmony with the democratic ideal. The creation, like a good government, was everywhere equable, benign, and evenhanded. The land was assumed to be democratic. That is, there was no such idea as ecosystem or habitat. The agrarian myth could not conceive of the aridity that was particular to the West. The lands in question were vacant not only of yeoman but of particular conditions and therefore particular knowledge, a blank slate needing only lines, plows, and bags of European seeds. Aboriginals were not nomads because of aridity, but because of ignorance.

Jefferson's attitude toward the Indians was one of fascination and respect consistent with the romantics' "noble savage." Nobility for aboriginals was the age's new wrinkle, but the notion of savage was an old one among civilized people. Jefferson, the champion of libraries and plows (he founded the Library of Congress, he held a patent on one type of plow), of the literate and agrarian, would of course see the nomads of the grasslands as barbarians in need of the farmer's beneficent care. His idea may have been softened by the appellation "noble," but later the modifier was easily dropped by others, who would wage a genocidal war along the same lines Jefferson had drawn.

Jefferson regarded Indians as educable but their life deficient. In 1802, he delivered this speech to a group of native leaders in Washington:

> We shall, with great pleasure, see your people become disposed to cultivate the earth, to raise herds of useful animals, and to spin and weave, for their food and clothing. These resources are certain; they never will disappoint you: while those of hunting may fail, and expose your women and children to the miseries of hunger and cold. We will with pleasure furnish you with implements for the most necessary arts, and with persons who may instruct you how to make use of them.

Jefferson hoped to use the Louisiana Purchase as a medium for growing yeomen, but also as a place where Indians could be kept in isolation until they were taught farming. Other than the depletion of their game, his vehicle for doing this was debt:

When they withdraw themselves to the culture of a small piece of land, they will perceive how useless to them are their extensive forests, and will be willing to pare them off from time to time in exchange for necessaries for their farms & families.

Donald Jackson points out:

A young Pennsylvania farmer, William Ewing, was sent to the Des Moines Rapids to teach the Indians agriculture. The irony was that the Sauk and Fox women, with their corn, bean and squash fields, knew more about farming than Ewing did.

William Least Heat Moon, in *PrairyErth*, speculates: "Surely, lore must have been deliberately withheld from a people taking away the land, so that the thieves got the big machine but not the operating instructions."

In his 1914 study of the ethnobotany of the upper Missouri region, Melvin R. Gilmore identified and documented the use of 169 species of plants by the Indians of that region. Among them were imports from Mexico (beans, squash, and as many as twenty varieties of corn), from Europe (melons were an immediate hit on Indian trade routes, probably because they were an obvious analog for the native squash), and tobacco, but the bulk of the botanical larder was indigenous. Gilmore wrote:

The people of the European race in coming into the New World have not really sought to make friends of the native population, or to make adequate use of the plants or the animals indigenous to this continent, but rather to exterminate everything found here and to supplant it with the plants and animals to which they are accustomed at home.

One of our early names for the American grassland grew from an 1820 expedition under Major Stephen Long. Edwin James, the survey's official chronicler, called the region between the Mississippi and Rockies a "dreary plain, wholly unfit for cultivation, and of course uninhabitable by a people depending upon agriculture for their subsistence." James's map labeled the area the "Great Desert," and later "The Great American Desert" became the favored and popular form of James's epithet.

The characterization was a biome too arid, of course, but it helps to understand that the words "desert" and "wilderness" often meant much the same thing in European and Christian mythology. In dichotomy, the word "wilderness" defined "civilization." Wilderness meant godlessness, which was why Jesus was dispatched to the wilderness for forty days as a test of the faith. Wilderness was in need of taming and deserts in need of being made to bloom. Thus, James's map warned off no one, at least not for long; on the contrary, it laid down the subtext that made settlement of the West a holy war of obligation.

The place was a mess, and it became a young nation's job to fix it with geometry, democracy, seeds, steam, steel, and water. The aboriginal information, the wisdom of the grasslands, was invisible to the agrarian and literate eyes of America. The philosophical way was cleared for the remaking of a place to a degree unprecedented in natural history. Yet seeing the events that would unfold as only exploitation, first by a naive agrarianism and then a rapacious industrialism, would be to discount the importance of what happened and what happens still. This was not so much a struggle between old ways and new ways or ignorance and knowledge as it was a battle between types of knowledge, inductive and deductive, between the wisdom of Jefferson's libraries and the received wisdom of the land in the library of plants among the Hidatsa, Arikara, and Sioux. Seeing this confrontation as only an exploitation by Eastern interests ignores the power of the empirical, the power of the land to teach and assert itself.

The rationalism of Jefferson's day was naive, even arrogant and elitist. It was a delusion that suggested all that is important about the world could be deduced in a drawing room. It was uniquely European and thoroughly Platonic in its idealism, its belief that ideas matter more than things. Still it was at bottom a faith in inquiry and science. Science is eventually self-correcting. Lewis and Clark collected specimens and made notes, and slowly these built back into the process. It was not expected, and it doesn't seem to be anywhere a motive of the early explorations, but eventually the wanderings in the plains built a national tradition of science. Credit this to the power of the land. It would take more than half a century, but eventually the rationalism of Jefferson would give rise to its antithesis in the character of John Wesley Powell.

The grassland region was the big unknown, and the focus then of the whole nation's attempts to know. National science was born in the arid West. The title of William Goetzmann's history of science as it evolved in the American West is *Exploration and Empire*, and this exactly lays out the issue and odd roots of nineteenth-century science across the nation, not just in the West.

The field on which this science's earliest inquiries played, however, was so vast as to intimidate attempts at explanation. This nineteenth-century science was not based in study of relationships among organisms. Instead of studying connections, a failing Gilmore was still lamenting in the early part of the twentieth century and a whole school of biology laments to this day that the early investigators were enumerators. They collected and counted the pieces of the puzzle, but made little attempt to fit them together. The inquiry had the feel of a banker inventorying the assets of an estate on which he had foreclosed. It was counting in preparation for empire. Writes Goetzmann of the work of Baird, Lawrence, and Cassin, the day's most celebrated biologists:

> Unlike similar ventures among whalers on the high seas, where the routes and migrations of the great mammals were carefully studied, no one concerned himself with the habits of the buffalo and other wild game. Even the beaver was largely ignored, as were the migratory fowl. In short, there was no attempt at all to relate animals and men on any level—either in the Darwinian realm of evolution or in the practical realm of human and animal ecology.

It is because of this penchant for collecting that today one can visit Monticello in Virginia and see the head of an elk taken by Lewis and Clark from near my home in Montana. The collections by Jefferson's envoys filled an entire museum in Philadelphia. The specimens from the range of expeditions filled the drawers and shelves of the Smithsonian, which was the originator of many of the early-day ventures and incubator of the national science.

Scientists collected the people of the West with the same vigor they directed to flora and fauna. Live people were rounded up and exhibited in Europe like circus animals; the scientists dismembered the dead. An assistant to the great geographer Alexander von Humboldt suggested pickling Indians and shipping them back in barrels. Army officers whacked up smallpox victims among the Blackfeet and sold their bones to the Smithsonian, bones the Blackfeet were able to reclaim only in 1989 and only after a considerable fight.

Through most of this, the nation accumulated very little hard information, and by mid-century, most of the grassland was still a great blank on national maps. Expansion leap-frogged the entire region and moved directly to California and the treed Pacific Northwest. The 1849 gold rush spurred one of the largest internal migrations of people in the history of the nation. Most of these people moved straight through the plains, but the place failed to capture their interests. Farmers settling the upper Mississippi pushed to the edge of the prairie, then stopped, believing that soil that would not support trees would not raise crops. They lacked steel plows to break the sod. The prairie offered no timber for houses and barns.

Then in 1837, a blacksmith named John Deere forged some significant improvements in the metal plow. During the same year, carpenters in Chicago perfected the technique of balloon framing, allowing houses to be made from light and therefore transportable frames instead of massive timbers. In 1868 John Lane, Jr., developed a three-layer plowshare and moldboard made with a center layer of soft steel that would not snap in prairie sod, as Mr. Deere's plow sometimes did. In the early sixties, Texas ranchers began moving their Mexican cattle north to supply Confederate troops. From this evolved the practice of sending cows farther north still, to the Ohio valley, to fatten on corn.

The Homestead Act came in 1862, the Timber Culture Act in 1873, the Desert Land Act in 1877, and the Timber and Stone Act in 1878. All were designed to parcel land to yeomen. President Lincoln, a direct ideological descendent of Jefferson's, signed the Homestead Act and made the railroad land grants of more than ninety million acres. The land was meant for the railroads to sell to settlers, providing them the inducement they needed to extend their lines.

Behind all of this, throbbing like a vein that would and did burst, was the national upheaval of the Civil War. It uprooted a generation of people and sent them wandering into open spaces. Those still in uniform when the war ended went looking for engagements and found them with the warriors of the plains. As enraged by war as the soldiers was the rising northern industrialism, ready now to ram its rails and smoke out across the plains. New plows, new lands, and upheaval; by the end of the Civil War, all was in place. The safety valve of vacant land Jefferson had envisioned a couple of generations earlier seemed necessary and was beginning to bleed live steam.

In the present-day West one may meet sun-bleached river rats, the summer version of the ski bum, who don't know or care to know history, and still these people will tell you that John Wesley Powell was an "awesome dude." He earns this praise because it became his task to chart the unknown reaches of the Colorado plateau by shooting his expedition down the whitewater of the Colorado River in four wooden boats in 1869. No one had ever done such a thing before and no one would think of doing such a thing today. The abyssal canyons of this most ferocious of rivers are today navigated only in gargantuan rubber rafts piloted by experts who know what lies around every bend. Powell did not. For all he knew, the next bend may have sent the river rolling off the edge of the earth. He had one arm, having lost the other in the Civil War, and he gained the rank of "major" in trade.

We know of Powell's work today not simply because of his river skills, which are shown to be lacking in the account of his journey. Rather, he looms largely for two reasons: First, the bulk of his contributions to the national knowledge following his great adventure was remarkable. Second, his work was properly recognized in a biography that was itself as prescient and remarkable as its subject. Wallace Stegner wrote *Beyond the Hundredth Meridian* in 1953, a full generation before a range of western writers would begin defining their efforts as bioregionalism, the notion that knowledge and custom are particular to place. Powell understood this a century earlier and Stegner recognized this understanding. Powell served notice that the natural conditions of the grassland called for a new set of abstractions, a new agriculture and literature defined by aridity.

By the mid-nineteenth century, the primary goal of science was to draw the dominant abstraction of the day on the land and complete the Jeffersonian formula of property, yeoman, and democracy. The land had to be diced up in rectangular parcels so that it might be owned to create freeholders as opposed to the free roamers who had inhabited the West before.

Accordingly, attention became drawn to surveyors, and mainly there were four: Clarence King, Lieutenant George M. Wheeler, Ferdinand Hayden, and Powell. Their mission, pushed and supported by the army, especially the work of Wheeler and King, was primarily mapping. The expeditions, however, generally included ornithologists, botanists, and zoologists, and therefore some interest in natural science.

There developed, too, tensions among the surveyors, because the enterprises were fodder for accounts in *Harper's Weekly, Scribner's Monthly,* and the rest of the press of the day. Celebrity was at stake, so competition arose, aided by the split between scientific and military goals. King and Powell became allies, as did Wheeler and Hayden, and for decades the two camps slugged it out.

The battle lines were more than personal and reverberated through Washington, the international scientific community, and the nation's attitude toward itself. Hayden, for instance, was a staunch ally of Edward Cope. King and Powell were tightly allied with Othniel Marsh, the fellow who found that horse in Nebraska and who also was a friend and colleague of Thomas Huxley's and Charles Darwin's. Thus the surveys became extensions of the international revolution in science, and the surveyors' enmities became tied up in the bickering between Cope and Marsh.

Far more was at stake in these government-sponsored surveys than maps. Writes Stegner:

> Out of Washington and its centralizing set of mind, as much as out of the West and the Western temper, came institutions that have shaped the West and to a lesser degree the whole country: Geological Survey, National Park Service, Forest Service, Coast and Geodetic Survey, Weather Bureau, Bureau of Standards, Bureau of Mines, Reclamation Service, many of them proliferating out of the mitotic cell of the Smithsonian. Government science before the Civil War was largely, though not exclusively, Joseph Henry and Spencer Baird of the Smithsonian. Geology was a states' rights matter, topography and mapping were diversions to occupy the peacetime Army, time and weather were for the Navy to play with, and too much of private science was the occupation of amateurs of the kind that Powell himself started out to be. Postwar Washington permitted and encouraged the development of professionals and put them in charge of operations of incalculable potential. Less than twenty years after the war, Washington was one of the great scientific centers of the world. It was so for a multitude of causes, but partly because

America had the virgin West for science to open, and Washington forged keys to open it with.

The initial result of this was forging those keys and then refusing to open the door. This great burst of creativity gave birth mostly to bureaucracies, but the science behind them was generally shrugged off when it came in direct conflict with the commercial urges that created the surveys in the first place. This is the story told mostly by the experiences of Powell.

The characteristic that most set Powell apart from his surveying colleagues was his lack of education. His science was self-taught, mostly gathered in the upper Midwest through constant youthful ramblings. Fresh out of the Civil War, he began teaching geology at Illinois Wesleyan University in Bloomington, lecturing in science and leading natural history tours, standing on not much more of a foundation than the bits and pieces of informal education he had picked up along the way. He used the university as a sort of springboard to exploration, mostly conning the institution and the Illinois legislature into sponsoring his early expeditions.

The act of climbing into the uncharted canyons of the Colorado River in a wooden boat was nothing if not naive, and this was Powell. As one might expect, the result was a series of disasters: three men died, although not on the river. Yet people still say in the West, as Nietzsche did, that what doesn't kill you makes you stronger, and from the Colorado Powell seems to have learned a healthy respect for hard information, as such respect can only be taught by an unforgiving terrain. It was this respect that led eventually to the *Report on the Lands of the Arid Regions of the United States*, finished in 1878. In a century full of nonsense about the grassland, it was the standout piece of wisdom.

After the Colorado voyages, the Powell surveys suddenly became distinguished from their counterparts by yet another factor: an intense focus. The others wandered hither and thither about the region, drawing lines and collecting specimens, but Powell settled down on the Colorado plateau for intensive study of that strange domain of yucca and slick-rock canyons. He also began relying heavily on local native populations. He was aided in this by a trail-hardened Mormon who had already learned that the best way to ensure one's survival in alien, parched terrain was to learn what the natives know. This exposure to the Ute, Paiute, Hopi, Shoshones, Moke, and others built in Powell a lifelong interest in ethnology, which he would insist on incorporating in all of the government science. At his death in 1902, by then stripped of all his other commands, Powell was still director of the National Bureau of Ethnology, which he created.

The intensity of his studies, aided as they were by his unique tie to local knowledge, gave Powell a unique view of the West, a view contrary to all of the nation's thinking about the place. Not simply a wilderness awaiting yeomen, the West was an environ distinct from the East and from the boreal reaches of Europe where our national weltanshauung had been incubated. To Powell, there was a single unifying attribute of the West: the place was arid. If there was to be successful colonization of the place, then patterns of settlement had to first and foremost respect this dictate of nature. This was the central theme of his 1878 report, and the notion that placed him at odds with the forces of his day and of ours.

He said settlement should be sparse, as it had been under the Indians, and should follow patterns already explored under Mormon occupation in Utah. These Mormon pioneers had learned something about the toughness of their land and Powell learned from them, studying early Mormon irrigation, how it took advantage of topography, and exactly what would be required to bring additional acreage under cultivation.

The popular doctrine of the day said that most of the West would simply sprout any seeds offered, but some of it would require irrigation, which could be installed "as easy as fencing, which it supersedes." After Powell's men watched the Mormons, they concluded that only 3 percent of the land area of Utah could be irrigated at any expense. The Mormons spent the seventy years following Powell's report applying every available ounce of technology to the task and by 1945 still had only 3.3 percent of Utah under irrigation. Forty-two years later, according to the 1987 Census of Agriculture, after two decades of intense development and emphasis on industrial agriculture in this country, Utah had just less than 3 percent of its area under irrigation.

The realities of irrigation led Powell to support a two-tiered system of agriculture, rather than the carpet of 160-acre rectangles offered by the Homestead Act and derived from the geometric subdivision of a square mile. First, Powell suggested that the intense labor and capital costs of irrigation meant the yeomen wouldn't be able to handle much more than about 80 acres. Parcels any larger would require wage earners, monopoly, and corporate farming, all patterns the official doctrine sought to avoid. That very small percentage of the whole West suitable for irrigation, then, should be divided into small and manageable plots of about 80 acres. The rest of the West, brought under agriculture but not irrigated, was destined for grazing, and thinking that anyone could survive on 160 acres of pasture in the West was surely insane. Powell proposed pasture farms each containing 2,560 acres, four full sections, sixteen times as large as the plot the Homestead Act specified.

His most interesting departure, however, was to fly straight in the face of the geometric perfection of the rectilinear cadastral survey. Powell had wandered the land, its rolls, meanders, and broad basins. He had a feel for its geology and how it moved and flowed, and suggested that settlement ought to do just the same. Rather than following the lines of abstraction, he suggested that the lines of settlement follow rivers and hydrological divides. This would ensure that every parcel took advantage of topography and in some way faced water. He proposed towns and villages fronting rivers, with intense farming areas around them. Lands farther from the river would be grazed.

Implicit in this notion was a challenge to traditional assumptions about property. That is, the right to own soil amounted to very little in the West. What would support and sustain the yeoman was not the dirt but the blessings of the land, its power. In the West, this life was not nearly so quantifiable or containable as to fall neatly into packages of 160 acres. It could not be bordered by the rectangles of the grid. In grassland, life flows, waxes, and wanes, upstream and down, up slope away from drought, toward refugia, escaping the rigid structures of geometry just as fluid dynamics escape the linear formulae of physics. The power of the land moves, and those who derive their power from the land must move with it.

Powell proposed a system modeled on the communal cooperation he had found among the Mormons and among the Mexican *ejidos,* the *metis* villages of the northern plains, and the Hidatsa and Arikara villages of the upper Missouri. He proposed a system consistent with the dictates of the land and derived from existing human culture, but not from existing agrarian culture.

Washington had a somewhat different idea. By the time of the release of his report, Powell's stature had grown considerably in Washington and the community that was waging national science. He had first firmed his standing with the vigorous backing of Joseph Henry, who was the first president of the National Academy of Sciences when Congress created it in 1863. Henry's successor, Othniel Marsh, was even more helpful to Powell and quickly sent the report first through a committee and then through the whole of the National Academy, resulting in a recommendation from the whole of the scientific community to Congress that Powell's ideas become national policy.

By then, though, the national interests had been set, and Western lawmakers united rather quickly in opposition to the egghead notions emanating from the National Academy. Certain powers in Washington chose to believe the boosterism of the day, some of it even sanctioned by official government theories, including the beliefs that the arid West was undermined everywhere by artesian wells that would immediately bubble to the surface under cultivation; that the root structure of plains plants was such that settlers would be able to dig for firewood; that three domestic cattle could pasture easily in the space used by one wild animal; that the area west of the Mississippi could support 1.8 billion people, according to reliable calculations; and the notion that rain follows the plow. This last hypothesis says that any form of cultivation would necessarily stimulate precipitation, an idea once championed by Hayden, Powell's colleague in surveying. Western senators labeled the opponents of these truths "scientific lobbyists" and "newly fledged collegiates."

Congress fought this issue to a stalemate and eventually worked to a strange compromise. That is, the sort of settlement pattern Powell envisioned would require a thorough knowledge of the West just to divide the land properly, and hence a consolidation and sophistication of surveys would be necessary. Congress went along with that and appointed Clarence King to head the newly created United States Geological Survey, a clear victory for the scientists. Powell, meanwhile, continued to rise in scientific circles, was appointed to head the Smithsonian's Bureau of Ethnology, succeeded King as director of USGS in 1880, and lasted there until Congress pulled his budget in 1894. His idea for democracy in the arid West, however, died. As part of the compromise, Congress killed Powell's prescient idea.

Evidence of its death need not come from accounts of the day, but is broadly written now and can be as easily read from any aircraft crossing the plains on a clear day. One must look hard for the natural contours of the land, raised only subtly by shadows. One must even look hard for the rivers, drained as they are. Their meandering lines are lost among the string-straight lines of canals that carry water to land-locked rectangular parcels.

But in flight, one need not look hard for Jefferson's grid. The wheat fields and cornfields lay out to the horizon like tiles. This is the obvious and squared face of the land that was once nothing more than an abstraction, a simple idea in the head of a man who never saw the place. All of the West is locked in a grid. The mask has become the face.

6

The End of Grass

The agrarian myth did not settle softly on the plain like a blanket of sound logic. It imposed itself by varying degrees of warfare, little wars among people but in the end wars against the land. The battles continue, prosecuted by every farm tractor that pulls a gang of field cultivators and a tank of anhydrous ammonia. It was and is a war against the life of the plains. First Indians lost to Indians, then all Indians lost to civilization's steeled army: bison lost to the railroads, then cowmen lost to the plows. It is this last skirmish that is our business just now.

During the latter half of the nineteenth century, the agrarians—rapidly becoming industrialized with their iron roads and iron plows—were fighting on three fronts. They were exterminating the bison and coincidentally most of the megafauna of the plains; they were exterminating or incarcerating on reservations the native people; and they were engaged in a literal shooting war with the cattlemen.

• •

Today, one can walk a federally managed wildlife refuge on the Missouri breaks of Montana, and although the carved wood sign at the entrance shows an elk, one is far more likely to see cattle on this land. In the federal Region 6, the bureaucratic subdivision that includes Montana, Colorado, and Wyoming, there are 109 federal wildlife refuges. Cattle graze on 103 of them. In southern Utah's desert washes, Anasazi potsherds lie shattered in prints of cow's hooves and spattered in cow shit. The Aldo Leopold Wilderness Area in New Mexico is acknowledged, by the federal bureaucrats who manage it, as severely damaged by grazing.

In Arizona, the Audubon Society bought an eight-thousand-acre ranch, free of cattle since 1968. Still the ranch waits for its grasslands to recover. The region's whole water table has been lowered by overgrazing, and the hydrology of the arid land will never be the same. Two hundred sixty million acres of U.S. Forest Service and Bureau of Land Management lands are leased to grazers at below market rates, but when the government began action in 1993 to increase grazing fees, a U.S. senator from Montana called it a federal "land grab"—of the federal land. In Montana, Wyoming, Idaho, Utah, Nevada, Colorado, Arizona, New Mexico, Texas, and even Oregon, one wears a Stetson to do business in the statehouse.

In total, 841 million acres, or 44 percent of the land base of the nation, is grazed by domestic sheep and cattle. Grazing is the single largest land use in the nation. In light of this, it seems strange to assert that the cattlemen lost that nineteenth-century shooting war with the plowmen, stranger still to assert that it would have been better if they had won, but indeed it would have been better. Free ranges and grazing are more in tune with the dictates of the land and the science of John Wesley Powell than what we have now: the legacy of the plowmen. In the sense of the land, barbarian as they might have been, those early-day cattlemen were right.

Willa Cather's poem "Spanish Johnny" arose from the grass of Nebraska, from the time when the plowmen beat out the cowmen there.

The old West, the old time,
The old wind singing through
The Red, red grass a thousand miles,
And, Spanish Johnny, you!
He'd sit beside the water-ditch
When all his herd was in,
And never mind a child but sing
To his mandolin.

The big stars, the blue night,
The moon-enchanted plain:
The olive man who never spoke,
But sang the songs of Spain.
His speech with men was wicked talk—
To hear it was a sin;
But those were golden things he said
To his mandolin.

The gold songs, the gold stars,
The world so golden then:
And the hand so tender to a child
Had killed so many men.
He died a hard death long ago
Before the Road came in;
The night before he swung, he sang
To his mandolin.

We shall take our time unwrapping the layers of this poem, important as it is to reading Cather, and important as the contradictions of Cather's work are to the story of the grassland. For now, though, let us remember that the olive man was trusted in the poem to hold wind, grass, and song. And for now, let us note that Cather capitalizes "Road."

• •

All cowboys to this day sing the songs of Spain, more than just a reminder of their inheritance of the original Spanish ponies. The wrap of a rope around a saddle horn is called a "dally," a corruption of the original Spanish term for the act, *dar la vuelta.* The McCarthy bit, the hackamore bit of the plains, began as the *mecate.* "Chaps," the name for leather leg covers worn in rough country, started as *chaparajos.* A Stetson is a truncated sombrero. Cattle ranching in North America was invented by the Spanish and Spanish-speaking peoples in northern Mexico, Texas, and California while they were still under Spanish control.

Spanish cattle were in the New World from the beginning, marching along with Coronado. Escaped animals proliferated, seeming to prefer the Southwest to their native Mediterranean hills. By 1555, vaqueros were rounding up bulls from the region that is now Texas for bullfights in Mexico. In 1598, colonists brought seven thousand head of cattle into what is now New Mexico to begin permanent settlement.

These colonies evolved under distinct institutions such as the *presidio,* or military outpost, the pueblo, and the mission. The mission developed the agricultural methods geared to the arid landscape and at the same time the rangy Spanish cattle that would become the basis of the early cattle industry. Later came the *ejido,* a system of land ownership or pattern of settlement. Like the *metis* settlements of southern Alberta, the *ejido* was a community in line with what Powell envisioned for the whole of the arid west, a central village located near water and surrounded by vast tracts for grazing.

The system worked its way into Texas and was there intact when the Treaty of Guadalupe Hidalgo split Texas from Mexico in 1848. Mexican cattle were driven north for market in Missouri as early as 1842. This system of export, of moving cattle, however, was not the main goal of the Mexican cattlemen. The *ejido* was primarily a community and a system for inhabiting the land. A few cattle were exported from the beginning, but they were mostly unruly bulls destined for the arena. Most of the cattle were raised simply to support the community that tended them. Meanwhile, the American colonies of the Eastern seaboard had developed a cattle industry of their own, largely on lands cleared of forest and seeded with imported grasses.

When I was a boy in the sixties, growing up in a farm community in northern Michigan, I learned early to spot a grass called timothy. Besides its somewhat personable name, it was distinctive, a delicate wire of a stem topped by a dense, fuzzy seed head, cylindrical and fat like a caterpillar. Timothy was good horse hay and a valuable find. Farmers regarded it as wild grass, wild like the woods near home and distinct from the cultivated brome grass and alfalfa hayfields. I grew up and left Michigan and forgot about timothy, until one day I spotted it growing along a mountain trail in the northern Rockies and wondered how it had come so far.

Timothy (*Phleum pratense*) takes its name from a colonial herdsman named Timothy Hanson who around 1700 spread the seed to New York, Maryland, Virginia, and North Carolina. By then the colonists were raising cattle and horses and needed grass. It was thought then that Hanson was spreading a New World gift, discovered first near Portsmouth, New Hampshire, by John Herd and known, before Timothy took over, as Herd's grass. The most significant name here, however, is probably Portsmouth, "port" meaning travel to Europe. Timothy is an Old World plant the English called "catstail grass," the most descriptive of all of its names.

By 1785, Thomas Jefferson would list timothy among the grasses of Virginia as if it were native. He listed bluegrass in the same entry, a species even better suited than timothy for tracing the spread of our ideas and our cattle. One associates the name "bluegrass" with Kentucky because it was originally thought native there. About the time of the American Revolution, settlers started moving into the Alleghenies and began the practice of burning cane brakes as a method of clearing pastures. When they did so, they found the brakes sprouted luxuriant mats of clover and bluegrass (*Poa pratensis*). The phenomenon, however, was testimony not to the grass's nativity, but to its mobility. Like timothy, bluegrass is probably native to England, although some speculate it came from southern Canada. It spread along trails and roads, an efficient enough traveler to beat the settlers into Kentucky. Today it grows on the remote roads and trails of Glacier National Park in Montana, where horsemen spread its seeds so that their pack stock might have something to eat. It supplants native vegetation.

In Kentucky, though, there was a meeting of sorts, one unnoticed, as botanical confrontations often are, but significant. Bluegrass met buffalo grass, a native species and close relative of the buffalo grass of the high plains. The colonial cows were beginning to work their way out of cultivated pastures and into the fingers of native prairie that widened to the grasslands beyond. In the early going these lands were considered useless for farming but good open range for cattle, if one could find a way to deliver the cows to people. In 1805, George and Felix Renick raised up a herd of range cattle in Ohio and drove them overland to market in Baltimore, and an industry began its motion, ready a generation later to merge with the Mexican herds from the south. Corn then enhanced the motion.

The settlers found that, with the aid of their metal plows, the prairie would raise that peculiar domestic tall grass: corn. From this there arose the practice of fattening grass-fed range cattle on corn and then walking this new tender, fat-marbled meat to urban markets. Those Mexican cattle showing up in Missouri by 1840 were being drawn into this loop. No longer raised for the sustenance of community, they were becoming a part of the specialization of the land. Cattle became commodity marching across the continent like a car through an assembly line ending at an urban showroom.

By the middle of the nineteenth century the descendants of the Spanish cattle had wandered and bred in the savannahs and deserts of northern Mexico long enough to become nearly wild beasts. Having evolved in the arid areas of Mediterranean Europe, they took to the New World's grasslands easily enough, becoming rangy and independent. The caballeros dealt with them seldom, generally at annual roundups when they would gather the meat a village needed from among the animals too inept to stay hidden in the sparse cover. Long-legged, the Spanish cattle stood tall enough for their bellies to clear the rough brush ground cover. Gaunt and muscular, they moved as adeptly as the rule of grassland required. Long-horned, they fended the challengers that came. They even developed an immunity to the ticks that plagued the region, a fact that would have much to do with western history.

We have come to call them longhorns. Even the inexperienced eye can see they are altogether different from northern European cattle. Even urban eyes that spot their relicts today fenced in a rodeo lot or some other roadside attraction can see that they are as different as a Pekingese lap dog is from its closest relative, the gray wolf.

The longhorns were ideal for open-range ranching as it evolved in Texas, and the early Texans knew this. Far from the pastoral husbandry of Europe and the Eastern seaboard, this early ranching was but a step removed from the buffalo hunting that had earlier sustained the plains. Instead of killing wild animals on the spot, the early cowboys gathered them on the hoof and drove them to market. The land they came from was unfenced and unowned, as were most of the cattle. The first herds were built taking advantage of the new political independence of Texas from Mexico. The Texans (and the Mexicans for that matter) developed the habit of international raids, slipping across the border in the same way plains tribes slipped into neighboring villages to steal horses.

Once the big herds were built, the proceeds moved north, first to Civil War troops and to corn-growing regions, and after the Civil War was over and after industrialization fostered urban markets, to famous railhead towns like Abilene, Fort Worth, and Dodge City. All of this coincided with the extermination of the bison from the plains: The railroads were in place, the markets had been built, and the prairies had been stripped of the cows' chief competitor.

About this time, as a rule during a series of London winters, the Thames River froze solid, something that has not occurred with any sort of regularity since. This is a part of the story, a period of recent history when the earth's climate cooled slightly. In the plains, cooler means wetter. There were periodic droughts through the mid-nineteenth century but generally it was wetter than normal, a condition that must have contributed to the optimism about the possibilities of settlement. This condition also favored the grass and it grew, especially when it had rested from the great herds of buffalo.

Rumors of grass belly high to a horse in Montana reached the Texas cattlemen, so more cattle moved. Herds of longhorns trailed north through the band of plains clear into Alberta. To a cowman, grass is like gold lying across the plains, and a free roaming cow is a miner. The cowman says even today, "Grass is worthless until a cow sticks its nose into it." As long as his cows were free to roam, there were fortunes to be made. By the 1880s, Montana cattlemen were turning profits of 25–40 percent.

While news of tall grass is not the sort that carries across the Atlantic, news of instant wealth often does. At first, the rumors were discounted in London and Edinburgh as Wild West hyperbole, but some evidence to the contrary had begun showing up on British tables.

The United States suffered a crash and depression in 1870 (the railroads had a hand in this, too), which temporarily deadened the American market for beef, by now a full-fledged industry. Of necessity, the cattlemen began experimenting with a new device, the refrigerator. In 1876, three million pounds of American beef crossed the Atlantic to England in refrigerated ships. By 1881, the cattlemen would be shipping a hundred million pounds a year. This caught British attention.

In 1879, Parliament decided it owed itself a look at the situation. Parliament sent investigators to Texas, who returned, according to the writer Mari Sandoz, "bug-eyed from this empire of grass. Their report started the British invasion from the Panhandle up the Plains to Canada." Both English and Scottish capital began flowing into the plains as rapidly as the new rails would carry it. It became fashionable in London to own a ranch and convenient at times to set up a prodigal son on a few million acres of western range while he grew past his errant years out of view of polite society. Box cars began shipping tea sets, brocade, and decanters across the plains, cattle on the hoof back.

By 1886, the U.S. Congress had heard a few rumors of its own it wanted to investigate. A formal inquiry found that twenty-nine foreign-owned companies controlled twenty-one million acres of western lands, most of it public lands. No one really owned ranches in those days. Ranchers simply occupied state-sized chunks of public lands. They didn't need land; they needed grass and the freedom to move on it. Congress found that the open spaces that had been reserved for honest yeomen had become the preserve of cattle barons—imperial and foreign cattle barons at that.

Economically the ranchers were successful but only to the extent that they capitalized on some longstanding rules of the land. Unlike the plowmen then pushing onto the eastern fringes of the plains, their habitation was not drawn from an agrarian abstraction but evolved as the longhorns had from three hundred years of successful ancestors. They did not channel their enterprise within the narrow confines of property ownership and the rectilinear cadastral survey. Their founding idea was motion. They flowed with the grass.

As with so many subsequent activities on the American grass, however, the industrialization of the motion was the beginning of its undoing. The motion was not contained within community, as it had been in the predecessor *ejidos*. It was international and began responding directly to international markets. It stepped across the limits imposed by place. The Indians did not exterminate the buffalo because there was no way to use them all within the community, no need to justify all the labor of slaughter. Likewise, the *ejidos* would not overstock the range; they had all the beef they needed.

London, Edinburgh, New York, and Philadelphia, however, will never have all the beef they need and would not hesitate to overstock the range. The general ebb and flow, the motion that had been the rule, became channeled into the roads that drained the power from the place.

In *Love Song to the Plains*, Sandoz recounts the following:

> "All the main streets in Nebraska lead east," a runner for a Boston investment firm wrote his lady friend back in Massachusetts in 1890. It was a provincial statement, an ironic half-truth of the Plains, the direction in which the wealth was drained, if not the one in which wheels still moved.

The migrations of the big herds from Texas north into Nebraska, Wyoming, Montana, and Colorado put the drovers at almost immediate odds with the yeomen already pushing into Kansas and Nebraska, especially Kansas. The near-wild beasts and near-wild men out of Texas cared not a bit about trampling crops or running off milk cows tethered to settlers' soddies. There were shootings, lynchings, and burnings, real enough occurrences that would later turn up in Hollywood's oaters.

More interesting in some ways, though, than the drawing of the battle line was the graying of it. The two ways of life churned along their borders to produce a sort of hybrid, and it was this hybrid that lasted, to the detriment of the plains. By now the fat, city-bound cattle were corn fed; no longer grassland beasts, they had become wholly dependent on the plowmen's corn. This set the stage for the next step, the elimination of those Mexican cattle.

It became very clear in the settlers' eyes that there was some mysterious force, a malevolence, in longhorned cattle. This was not imagination. Not long after a herd of longhorns would work its way through Kansas, the blooded cattle of the settlers, milk cows, and working oxen drawn from northern European stock would begin dropping dead like a clear-cut bovine forest. The settlers blamed the Texans and gunfire erupted. Whole states banned cattle drives, an act the Texans took as a nasty prejudice against their animals, which obviously could not be diseased. It was, after all, the settlers' shorthorns that were dying; the longhorns were fine.

The longhorns had spent three centuries on the grasslands learning the place's rules, which included building an immunity to the fever carried by ticks. The shorthorns had no such immunity and died in the wake of the spread of the ticks during the cattle drives. One can imagine a similar sort of event when the invading Asiatic mammals moved across the Bering land bridge. In this later case, though, man had intervened and imposed rules that began working against natural selection.

By 1883, the cattle moving into Montana, for instance, were coming not so much from Texas as they were from Oregon, Florida, Ohio, and New York. The shorthorn was moving in, aided now by the British influence, the settlers' anger over tick fever, and a taste for "blooded" cattle, as the northern European stock was known. The difference was more than appearance.

The shorthorns evolved mostly around the British Isles, a fact still apparent from their names, Hereford and Aberdeen Angus. Standing alongside a longhorn, the northern European cattle look like sumo wrestlers next to marathoners. Sluggish and sedentary, they define the term "bovine." They could be handled more easily than the longhorns, and they took to the by-now standard practice of fattening on corn. Their fatty, tender meat suited the bourgeois demand growing in Europe and the cities of the Atlantic coast. They were not, however, in any way suited to the American West they have come to symbolize.

Shorthorn cattle evolved in a wet climate. They developed those soft, fleshy bodies grazing selectively to plants called forbs, the leafy plants, as opposed to grasses. In the West, they ate grass because the grass was everywhere, but so, at least once, were the forbs. The cows bunched up in the West, sorting themselves along stream banks and river bottoms where there was enough water to produce a flourish of green, leafy plants. It was a selectivity the West had not known, at least not known on a vast scale, in its twenty-five million years of grazing under the teeth of hoofed animals ranging from camels and elephants to bison. The bison ranged the highlands, ate grass, and moved on. The shorthorn cattle bunched up in the coulees, streambeds, and draws and grazed the land to dirt. In the century since, they have rearranged the face of 40 percent of the nation's land. They were helped in this by an invention that appeared in 1874, the barbed-wire fence.

A part of the Western myth that is true is the early-day cowboy's hatred of fences. Those first fences meant settlers. They were the settlers' tool for fending off marauding herds of cattle and defending their homesteaded quarter sections against the cattle barons. Anything unfenced was open range, and so in a sense the hatred of fences was simply a way of saying the cattlemen wanted everything; the hatred was greed. This may have been so, but it must be understood that the land-owning rancher was the exception. The closest analog of the early cattlemen was the nomadic herdsmen of the world's other great grasslands, the Mongolian steppes, the pampas of South America, and the Serengeti. Their wandering was simply a way of letting their herds wander with ever-changing grass. This had been the rule of the plains since long before Europeans could imagine rectangles.

In fighting the fences they were really no different than a Mongolian prince fending off the irrigating Chinese horde. But in another way they were very different. Before the roads and the rails, such a war would have been fought ultimately for a freedom of motion that was simply freedom. Subtly, though, our technology had begun to redefine freedom as the motion of capital, not people. Freedom had become the ability to exploit to benefit distant princes.

At bottom, the Johnson County cattle war was about fences, the literal lines the social forces drew. But when the Wyoming Stockgrowers Association met in 1891, presumably to vote an act of war, there were forty-three members present, only eight of whom came from ranches owned by individuals, the rest being owned by corporations, mostly British. Sandoz says pitching this group against the yeomen settlers, who were labeled by the cattlemen and subsequently by Hollywood as the "rustlers," drew a line in the sand that scratched to the core of the nation's definition of itself:

> Papers from Omaha to New York made a great story of the wrongs suffered by the cattlemen, perhaps not understanding the meaning of the term "public domain" or willing to side-jump it if they did. Very few understood that even the actual rustler was no more than an excuse for what really was a cattle war—a war of the ranch interests against the government and its avowed public-land policy: free land for everybody, a 160-acre place for every bona-fide homeseeker. This was the essence of what America meant to the peoples of the earth and against this the cattlemen were warring, the crusade being joined.

This particular chapter of the crusade became known as the Johnson County cattle war, but was by no means unique to the period. The troubles began almost immediately with the incursions of the Texans into the north, fueled occasionally by some lasting enmity from the Civil War and a Western taste for alcohol and violence, but at bottom, most of the skirmishes were about the free range. Settlers were shot, their shacks and corn burned. They were lynched, burned, lynched and then burned. Sandoz's own father, a land locator for the settlers, was a somewhat notorious combatant, having achieved some renown with the precise rifles he fired. His battles in the Sandhills centered on the Spade Ranch, a free-range spread.

In Montana, the legendary baron Granville Stuart (the mention of his name still produces hushed tones and doffed white Stetsons at Cattlemen's Association barbecues) presided over a social club known as the Stranglers. In July of 1884, they attacked a cabin on the Missouri River and burned it to the ground, leaving five charred bodies inside. The ranchers rode down and hanged anywhere from nineteen to seventy-five men who escaped the attack, depending on who tells the story.

In New Mexico, General John A. Logan retired from his military career to a cattle empire and hired one of his former soldiers to assassinate settlers. Logan became a U.S. senator and running mate to James G. Blaine on the Democratic presidential ticket of 1884. The violence occurred throughout the West, but in Wyoming's Johnson County, it came to look more like a war, extending into the governor's office, the National Guard, federal garrisons, and the Congress.

The epicenter of this was the town of Buffalo, lying then and now tight against the sweep of the Bighorn Mountains, which wall the high plains' westernmost edge. The war itself actually stemmed from an incident in nearby Sweetwater County, where two settlers, a man named Jim Averill and a single woman who was his neighbor, Ella Watson, were scratching for the promise in the Homestead Act. Averill was a doctrinaire subscriber to the American dream and a vocal critic of the cattle barons as the antithesis of that dream. Watson, meanwhile, was making some forays into a small-scale cattle operation. It is said she entertained cowboys or did their laundry or both, but they tended to pay in cattle, mavericks rounded from the range. She had amassed a few head, maybe forty. This gave her the name Cattle Kate and the need to register a brand.

Enforcement of the brand was the raison d'être of the cattle associations and the mechanism for enforcing monopoly. Because the range was open, the cattlemen would conduct a communal roundup each year, separating the cattle according to registered brands. If a cow was unbranded or wore a brand not recognized by the association, it was a maverick. Mavericks were either divided among the registered brand holders or sold to pay for certain business activities of the association, such as acquiring guns and ammunition.

Cattle Kate's owning cows and her attempt to register the brand was as great an affront to the ranchers as were Averill's ravings about the American dream. Averill had the unfortunate habit of publicly criticizing the ranchers in letters to the editor of the local newspapers. One day a delegation of ranchers kicked in the cabin doors of both Cattle Kate and Averill and hanged them. In full roar, the ranchers then kicked in the door of a settler who happened to know something about revolvers, and he staved them off, poking a few holes in the opposition. This act of defiance brewed to a general state of unrest all up and down the Bighorns and eventually led to that meeting attended by the forty-three members of the cattle association. We can only presume it was a council of war, because no minutes were kept, but the public record does show that the next day association money bought up every available rifle in the area. The governor himself gave the association a case of rifles from the state's arsenal.

The cattlemen had already taken steps to raise a small army. They sent a man named Ijam and the legendary regulator Tom Horn (who would later hang in the Missouri breaks region for killing a settler boy) on recruiting trips in Idaho, Montana, and points south. Two days later, a six-car train made the run from Denver to Cheyenne, bearing the proceeds of the effort. Sandoz describes the entourage in detail and it is worth repeating her account here to show the social weave of the cattlemen:

> Three [of the six cars] were full of horses, bought and branded in Colorado by R. S. Van Tassel, one of the main backers of the Invasion. For a while, he had been a partner of Tom Swan, and the son-in-law of Alex Swan, whose empire was lost to [Rancher John] Clay and his financiers. Another car held the equipment and the flatcar was piled with the three new wagons and camping stuff.
>
> The blinds of the one passenger car were kept drawn close, but rumors flew like gray snow on the April wind. Some realized that this was the start of the long-planned invasion of Johnson County. Inside were Tom Smith's twenty-five Texas gunmen, with their saddles and ammunition, and two adventurers who had attached themselves to the group, a young English rancher from Colorado and Dr. Charles Penrose, younger brother of the

Pennsylvania senator and the Colorado mining engineer, perhaps induced to come because he had been the medical classmate of the acting governor of Wyoming, Dr. Barber. Two newspapermen came to join, and the one recruit that Ijam's trip to Montana and Idaho produced, nobody from Tom Horn's search of the Dakotas. Twenty-four Wyoming regulators got on at Cheyenne, most of them in great contrast to the rough-and-ready Texas gunmen—many of them northerners well educated and cultured and cosmopolitan. Two were the owners of the Duck Bar Ranch in which Theodore Roosevelt was said to have an interest—the elegant Teschie, H. E. Teschemacher, Harvard bred, a world traveler, his parents living in Paris, and his polished and handsome partner, Fred DeBillier, owner of a villa in France. There was W. J. Clarke, state water commissioner and rancher of the Crazy Woman Fork; J. N. Tisdale, state senator and his son Bob, both with ranches up north; A. B. Clark, owner of the DE; the Englishman R. M. Allen, general manager of the British-owned Standard Cattle Company over at Belle Fourche not far from the Waggoner hanging and tied in with Baxter; J. C. Johnson, the eccentric partner of Tom Sun who had been at the lynching of Averill and Kate; C. A. Campbell, a Scotch Canadian who lost his cattle in the 1886–87 winter and went in with the Clay-Robinson Company; Bully Booker under Clay in the Swan Ranches; A. D. Adams, the burly Scotsman who managed the Ferguson Land Company and today delegated to ride close to Pap, W. E. Whitcomb, the gray-haired old cowman, in the country from back in the beaver-and-buffalo-robe days and much too wise to be riding the stormy trail to Johnson County today.

This ad hoc army carried what became known as the "dead list," the names of seventy men the cattlemen labeled as rustlers who were to be killed. Among the names were Red Angus, the sheriff of Buffalo, the town's mayor, the Johnson County commissioners, businessmen, and other officials. By the end of the first day, two of them were indeed dead, including Nate Champion, the settler who earlier had successfully fended off an attack. But those killings took place in a siege that was broken by firing a cabin, leaving badly burned bodies, and hearing this inflamed the settlers, who then raised their own army, pinned down the cattlemen at one of the ranches, and prepared their own fires. The siege brought federal troops, who arrested but in effect rescued the ranchers. The ranchers wound up in Cheyenne, supposedly jailed but really in custody of their attorneys. A jury would not convict them for the two killings or the wave of violence around the Bighorns. No one did prison time.

The range wars would continue throughout the grassland until the turn of the century. Several times the federal troops would ride on ranchers trespassing the public domain. And several times the ranchers would ride away from courtrooms free or having paid minuscule fines. By 1904, the cattlemen had already appropriated the barbed wire of the settlers and begun using it in a curious fashion: to fence the public domain. The practice was illegal, and in 1904 a president, himself a rancher and one the cattlemen thought they could count on to protect his western colleagues, began what were known as the "Roosevelt Roundups." Federal marshals arrested ranchers from New Mexico to the Dakotas, including some involved in the Johnson County wars. The last group was fined five hundred dollars apiece for appropriating the public's lands, and spent a day in the custody of the sheriff.

In November of 1905, two cattle barons were charged in Nebraska with trespassing on public lands and obliterating survey markers, which proved their crime, a war on the lines of civilization. Government surveyors had spent the previous summer working under the constant protection of the Secret Service to build the case against the two. Faced with this evidence, the ranchers pleaded guilty to the fencing of 212,000 acres of public land, for which they were fined three hundred dollars apiece, and spent six hours in the custody of their attorneys.

Roosevelt was enraged and some heads rolled in the federal justice system. There were new charges filed and the prosecution of trespassers spread, as did the use of federal troops to enforce the law. The day of the free range ended. In reality, though, this was something of a symbolic ending; the free range had died a generation before, and had really never lived as a stable economy. The cattle wars were an admission that the system was already on the ropes.

From the beginning, as the evidence of the boom mentality behind the spread of the industry through the West, sections of the range were fiercely overgrazed, and there were records of the "big die-ups." The cattle had stripped the range of the forage that could have carried smaller herds through the winter. More to the point, though, the massive winter kills of cattle were simply evidence that cattle are not suited to the periodic rages of plains weather. Cyclical drought and occasional severe winters are a way of life in grassland, conditions the bawling European cattle were not evolved to accept. And they were conditions the European and New York financial and commodities markets that had evolved were not willing to accept.

When the snow lies deep and hard-crusted across the plains, it only barely hides a rich world below. The native grasses cure well and carry their nutritional value through the winter. Bison have huge flat heads that they swing from side to side when grazing, like snow shovels, plowing a swath to the grass. Cattle stand and starve, their noses a few inches above cured carbohydrates.

In the winter of 1885–86, the ranches of the high plains showed losses of 50–75 percent. Sandoz speaks of creek beds filled solid with cattle carcasses for ten-mile stretches. A single ranch in Nebraska lost a hundred thousand head of cattle that winter. "The spring roundup of 1887 was the darkest, grimmest ever known to cattle country," writes Sandoz. In Montana, Granville Stuart lost twenty million dollars, and Conrad Kohrs salvaged only three thousand head from thirty-five thousand he owned that previous fall.

The cycle repeated itself in 1906–7, especially in Montana. Wallace Stegner was born shortly thereafter and raised near the Cypress Hills around the cowboys of southern Alberta and northern Montana. He says those disaster years closed a way of life:

> The net effect of the winter of 1906–07 was to make stock farmers out of ranchers. Almost as suddenly as the disappearance of the buffalo, it changed the way of life of the region. A great event, it had the force in the history of the Cypress Hills country that a defeat in war has upon a nation.

The ranchers began making hay to winter-feed the cattle. They fenced river-bottom fields, bought equipment, and raised crops of grass, first native, then imported exotic grasses mixed with alfalfa (an exotic forb) to put up for winter feed. It is a practice that is ubiquitous in the West today. The free range of western myth existed only as myth. The limitations of the shorthorn cattle quickly made the ranchers' work look like that of the settlers, horses or no. The section line fences went up, the crops went in. For twenty-five million years the grasslands of the West had been reshaping animals to fit its harsh dictates. European settlement, especially that by northern Europeans, brought the extermination of these animals and a reshaping of the land and its plant communities to accommodate the animals the Europeans imported. In a sense, this process succeeded because the land has indeed been reshaped, and in a sense it failed because the rebuilt landscape cannot sustain the new way of life. The design is shortsighted and faulty.

The West lives strongest and purest in myth. Stetson hats and horsemen's boots aside, the hay balers and fence builders of today's West, pushing their dogies along with motor bikes, have more in common with the yeomen than with the vaqueros the myth celebrates. And still the British bovids are in charge.

Nordic colonization of the New World, at least in its first wave, lasted at least five hundred years. Viking colonists were in Greenland one thousand years ago. They imported cattle. There are indications the cattle made it as far as the new place the Vikings called Vinland (after North America's native grape vines, the species that gave us the Concord). Unlike the Spanish cattle with Coronado, though, the animals with the Vikings could not survive in eastern Canada after they were abandoned by the colonists.

About 1500, the world's climate began a slight cooling. The name Greenland was appropriate when the Vikings found it, but it became, as it is now, mostly covered in ice and snow. The onset of the ice threatened the cattle-based economy of the Vikings. There were alternatives. During the five hundred years of settlement, the colonists had been in sustained contact with the Inuit people thereabouts, themselves a maritime people, who hunted whales, seals, and fish. This way of life was open to the Norse if they would learn it, but they wouldn't. They refused to abandon their cows and lost their hold on Greenland in the face of the advancing deep winter. Such is one of our ancestor culture's attachment to cows.

In 1991, Ron Stellingwerf, a district ranger for the U.S. Forest Service, one of two federal agencies that oversee grazing on public lands, made known his intentions to assess the environmental damage from grazing on the Beaverhead National Forest of Montana. Shortly thereafter, he found scribbled on his door the measurements of a grave, an antique vigilante warning. Stellingwerf moved to another town, and cattle still graze that section of the upper Ruby River, a trout stream in the headwaters of the Missouri, without undue governmental restriction on government land.

On a central stretch of Idaho's Snake River plain, another Forest Service ranger has been somewhat more intransigent. In 1987, Don Oman visited a Bureau of Land Management demonstration project designed to improve range conditions and saw the possibilities for the land in his care in Idaho's Sawtooth National Forest. He began making those changes, but ranchers whose allotments were reduced began pressuring the Forest Service through Idaho's congressional delegation. In 1989, the Forest Service quietly made plans to remove Oman, although there was no official criticism of the soundness of his management plan. Oman filed a whistle-blower complaint and the agency backed down. In 1990, *The New York Times* quoted a rancher as threatening to kill Oman. Again the Forest Service said the proper solution to all this was to assign Oman a job elsewhere.

George Wuerthner is a former agency scientist, a man of the grass with a graduate degree in range management. Mostly, though, he is mad as hell, especially if one mentions cows. He lives in Livingston, Montana, on the north edge of the greater Yellowstone ecosystem, a wildland including the park and the ring of national forests around it. This is a place known for its bison, elk, and grizzly bear, but even here the cows outnumber all of these wild animals put together, and this makes Wuerthner mad. His anger will ripple in long interviews, the sort that spin from a single question into a two-hour weave of dismal statistics.

Wuerthner quit the BLM and makes his living now as a free-lance writer and wildlife photographer, but his passion is speaking against grazing. At lunch, he scowls a bit when the philistine writer orders a cheeseburger; Wuerthner eats no beef. He talks about his plans to leave Livingston soon, although he and his wildlife biologist wife love the northern Rockies, especially here on the north edge of what is appropriately named the Paradise Valley. The politics have become too threatening.

After lunch, we walk up a hill at the edge of town. He says I have to see it. It's the sort of hill every western town has that nobody quite owns and so nobody has figured out what to do with it, although subdividers are changing that fast. This sort of plot is small enough to be forgotten—that is, to be unafflicted with cows, which is why we are here. The blue bunch wheat grass, this region's dominant native, grows in thick and uncropped tufts here and probably nowhere else like this in any of the grand valley that runs to the east for miles until it breaks against the Crazy Mountains. Wuerthner comes to this spot for the reason a lot of us do: to take a quick feel of the native grasses and the delicate flowers around them, and then look out all across the broad valleys and imagine it as it once must have been: a broad, tawny sea of grass, of bison and elk.

Slowly, the West was divided, fenced, and squared, but not all in the quarter sections that were to be the yeomen's havens. Even from the very beginning, some of the ranchers, led especially by the owners of the legendary King Ranch in Texas, believed the future lay not in longhorns and free range but in deeded land and blooded cattle. And almost from the beginning, some of the yeomen understood that their futures could not be contained in 160 acres of oats, but in full sections and double sections, acquired patiently through the years as neighbors died or went broke. Some of that land became stocked with a few head of cattle. There was convergence between ranchers and settlers.

The ranchers began buying up the larger tracts, especially railroad lands not inhibited by the strictures of the Homestead Act. They used dummy entrymen, hiring people to file on homesteads, and added the quarter sections of public lands to their heaps. The holdings of deeded land grew, but always as a sort of home base, not a limit. Their cattle continued to range from the home base to the still-public domain as they had in the 1890s. International markets built, the economy boomed, and sheep and cattle grazed 40 percent of the nation's surface to dirt. The cattlemen fought one another and the sheepmen for the remaining grass.

By the early thirties, the dirty thirties, the degradation of the range had become so widespread that even western senators crusaded for reform. The result was landmark legislation, the Taylor Grazing Act of 1934. From this arose the system of grazing allotments that we have today. Under the system a rancher is given a right to graze a certain piece of publicly owned land. Range managers, working either for the Forest Service or Bureau of Land Management, determine how many cattle or sheep a piece of land will support, a division called an Animal Unit Month. In western Montana, for instance, an AUM is about 20 acres, the amount of land needed to grow enough grass to support one cow and a calf in an average summer's grazing season. The rancher is allowed to stock his range at the determined level and pays the government rent on his allotment, as of now, $1.86 per AUM. (On private lands, the fee is typically about $10 per AUM.) These reforms were to end the abuse of the range.

Two years after the passage of the Taylor Grazing Act, the Forest Service reported the results of the nation's first comprehensive survey of the range. Scientists had examined more than twenty thousand plots throughout the grassland, comparing grazed areas with relict patches in cemeteries, along railroad tracks, and in out-of-the-way places. The survey found the range's carrying capacity —its ability to produce forage—had been depleted by 52 percent. Thirteen percent of the West had suffered "moderate depletion" (0–25 percent loss of forage); 35 percent, "material depletion" (26–50 percent loss); 36 percent, "severe depletion" (50–75 percent loss); and 16 percent, "extreme depletion" (76–100 percent loss). Put another way, more than half of the range had lost more than half of its plant life. It is important to note that by loss, the survey did not mean a percentage of the grass had been eaten; it meant a percentage of the plants that produced that grass had been eradicated by overgrazing.

"The plant cover in every range type is depleted to an alarming degree. Palatable plants are being replaced by unpalatable ones. Worthless and obnoxious weeds from foreign countries are invading every type and throughout the entire western range the vegetation has been thinned out until even conservative estimates place the forage value at less than half of what it was a century ago," the survey said. "There is perhaps no darker chapter nor greater tragedy in the history of land occupancy and use in the United States than the story of the western range."

That dust bowl era survey is the baseline, a snapshot of "before" outlining the charge of the Taylor Grazing Act. The act itself was an admission of those conditions and an attempt to control a half century's worth of overgrazing on the plains. It was scientific and progressive, and therein lies the fault. The progressives' notions of the early twentieth century were nothing more than extensions of the rationalism of the eighteenth, an extension of our mulish insistence that nature sign a contract to produce what we want. The stocking levels the scientific range managers decided on under the new act, the levels that were supposed to bring progress to the range, were based on an average, a mean, an expression of human and capital's need for predictability and stability. In grassland, however, there is no such thing as an average year.

"I've been here fifty years and last year was the first normal year we've had," an old-timer in Arizona once said to me. Arithmetically, the wild swings of drought, wind, and winter do yield something of an average, but don't expect to see it. An average stocking level during a drought year is a disaster from which the grasses cannot recover. Subsequent surveys of the condition of the range, then, should come as no surprise, and this is what makes Wuerthner so mad.

The federal government's last comprehensive survey of the range, which was done in 1980, listed only 15 percent of the range in "good" condition. Sixteen percent was "very poor," 38 percent "poor," and 31 percent was listed as "fair." Interestingly, the government gave itself its widest window in deciding its definition of "good"; a piece of ground needed to have 60–100 percent of its potential plants. "Very poor" land had less than 20 percent of potential. The survey concludes that 85 percent of the nation's grasslands have less than 60 percent of the potential vegetation, all as a result of overgrazing.

Says a separate report commissioned by the Environmental Protection Agency: "Extensive field observations in the late 1980s suggest riparian areas throughout much of the West were in the worst condition in history."

A 1989 survey of only BLM lands showed that 68 percent of

those—more than 100 million acres, an area the size of all of New England, Pennsylvania, New Jersey, and New York combined—is in "unsatisfactory" condition. This is not simply a problem of mismanagement or exploitation of government land. In 1987, the Soil Conservation Service estimated that 64 percent of all privately owned range land is in unsatisfactory condition.

Wuerthner had been sitting for the interview as we waded into these numbers, but he stood up then and began pacing the room. "Unsatisfactory" in the last two surveys means that more than 50 percent of the plants—not volume, not numbers of individuals, but 50 percent of the species one would expect on a pre-settlement grassland—is absent.

The bottom line is that on the greater part of the range land grazing exterminates at least half of the species. Cows are selective in what they eat; they concentrate their lives, their manure, and their grinding hooves in riparian areas. Because of all of this range managers have come up with the notion of "increasers" and "decreasers," based on which plants thrive on the disturbance of grazing and which wither. The increasers are by and large a specialized bunch we call weeds, especially exotic invaders. The decreasers are all the rest of the species, more than half.

Wuerthner, whose job it once was to perform these surveys, says: " 'Good condition range land,' this is another thing that is one of their euphemisms. If you know how these range conditions are determined you wouldn't call it good. They do a comparison. They look at the site. 'What plants could we expect to grow here.' If seventy-five percent or higher are there it's 'excellent.' If it's fifty percent, just more than half, we call it 'good.' Suppose I went out in the forest here and said that because of some logging you could no longer find any ponderosa pine and Douglas fir. Would we say the forest is in good condition then? We'd be outraged."

Blue bunch wheat grass, the mainstay, the capstone of the shortgrass prairie along the northern Rockies, is a decreaser. Grown out of reach of cows, it sprouts to a head in August, spreads a tawny fanned tuft of leaves and stem on the landscape, then stands into winter, carrying its stored nutrition with it. The elk and deer leave it alone until winter, when they dig it from the snow like a farmer takes potatoes from his root cellar. Grazed in winter, it feels nothing. Cows eat blue bunch during the growing season. Cropped in summer, it withers, sacrificing the root-building process to build new leaves. Sacrificing roots is sacrificing the underground option, the necessity of survival in drought land. Recent research shows that a clump of blue bunch cropped during the growing season takes ten years to recover.

In an arid land, cows bunch up along water. State water quality officials say there are 165,000 miles of degraded streams in the nation and 64 percent of those have been degraded by agriculture, mostly grazing. A report by the federal Government Accounting Office in 1980 said 80 percent of the riparian habitat managed by the BLM in Idaho is "degraded"; in Nevada, 93 percent in one area of the state, 86 percent in another; and in the Tonto National Forest in Arizona, 80–90 percent.

Gentle streams meander and push through soft grassland channels hidden by willows and brush. The roots support the streambed and form mini-dams that slow the water to produce periodic floods. These water the entire riparian zone, which sometimes stretches out several hundred feet on each side of the stream. Cows eat the willows. In recent years in a region of Nevada where the cows weren't doing a thorough enough job, the ranchers poisoned the willows with herbicide on the theory that the brush drinks water and water is for cows. Stripped of the diversion of streamside vegetation, the water speeds up and starts to down-cut, eroding to a new and lower bed. Once the water level is thus lowered, the willows can't come back, even if the cows are pulled off. Their roots can't reach the water. The land is forever changed. The water is forever siltier, faster, and warmer.

The West is veined with dry washes where there once were streams and brooks. Removing the cows from a five-mile stretch of a grassland creek in Nevada increased stream flow 50 percent.

Of the 150 fish species native to the West, 122 are either extinct, listed as endangered, or are candidates for listing. Cows are implicated in virtually all of the cases. The nation has spent billions damming the West, digging canals, and stopping up rivers behind reservoirs on federal subsidy. Thirty-seven percent of all of this water goes to the production of livestock. In California, the nation's most populous and urbanized state, where city and suburb dwellers don't flush their toilets and water shrubbery only on certain days to conserve, water for livestock outstrips all municipal uses of water.

The Midwest is covered not with cows, but with corn and other grains, but 70 percent of this grain goes to feed livestock.

A cow produces sixteen times the feces of a human. The number of cows grazing near and on the mountain streams of the greater Yellowstone ecosystem produce the equivalent sewage of a city of 1.6 million people. Most of it washes into those streams, all of it untreated.

At times, even Yellowstone National Park appears overgrazed. There are no cattle there, though its grassy meadows are cropped to the nubs by elk and bison. Yet exclosures, which fence grazers out of a small area in the park, show no difference in species diversity and the overall health between grazed and ungrazed plots. Exclosures of cattle, especially in riparian areas, show drastic differences. Even on the uplands away from streams, exclosures in the Great Basin yield land of about 10 percent sagebrush and the rest native grasses. On land grazed by cattle, those proportions are often reversed. Big sagebrush is an increaser.

In 1915, urged on by cattlemen and sheepmen, the federal government began "predator control" of the western range, an effort that deliberately extirpated wolves in the West. Between 1916 and 1928, federal, state, and grazing association agents killed 63,145 animals, including 169 bears, 1,524 bobcats, 36,242 coyotes, and 18 mountain lions, along with uncounted badgers, beavers, civet cats, blackfooted ferrets, foxes, martens, minks, muskrats, opossums, raccoons, skunks, weasels, porcupines, rattlesnakes, prairie dogs, ground squirrels, jack rabbits, eagles, and magpies. The killing has not stopped, but has been subsumed in a Department of Interior program called Animal Damage Control, which in 1992 was budgeted at $35 million. That money pays for 835 federal employees whose job it is to kill or otherwise harass wildlife, mostly for the benefit of the livestock industry. During the federal fiscal year of 1992, the agency claimed to have killed 22 million birds, mammals, and reptiles. In one case in New Mexico, the agency spent eighty staff person days using snares, traps, and sodium cyanide baits to kill thirty-three "target animals" and twenty-two "non-target" animals, a rampage provoked by the loss of one lamb.

The average Montana rancher has a net worth of $800,000. A 1993 study by the federal Government Accounting Office said that 6 percent of the ranchers with permits to graze federal lands accounts for almost half of all federal grazing. That is, the federal subsidy in cheap grazing fees benefits the cattle baron, not the family farm. *The Washington Post* reported that among the largest is the Sieben Ranch near Helena, Montana, the ranch of the family of U.S. Senator Max Baucus. Other major grazers are the Hunt Oil Company of Dallas, Texas; the Federal Mutual Insurance Company; Metropolitan Life Insurance Company; a Japanese conglomerate; the Mormon church; and the Church Universal and Triumphant. A single cattleman, Daniel H. Russel of California, holds permits on five million acres in California and Nevada, an area the size of the state of Massachusetts. The five hundred largest grazing allotments average 58,178 acres apiece of federal, public, and degraded lands. All of the grass on all federal lands produces about 3 percent of the forage for all of the nation's beef.

In 1850, with 100 percent of the western range still in excellent condition, with all of the plant and animal species present—and without benefit of fences, federal subsidies and their attendant lobbyists, hay balers, reservoirs and canals, cattle trucks and refrigerated cargo ships, predator control, antibiotics, public hearings, squeeze chutes, hormones, frozen semen, and Kawasaki quarter horses—there were about ten million elk and maybe as many as seventy million bison, but probably closer to fifty million, on the Great Plains, according to the most reliable estimate. Now a century and a half later, elk, bison, antelope, and deer have been reduced to about 1 percent of their original numbers in the West, replaced with cows—specifically, 45.5 million head of cattle in the ten states that were the main area of the original bison range, according to the 1987 Census of Agriculture. A cow produces about as much meat as a bison, but it is fattier meat, much higher in cholesterol. Many who eat both prefer the taste of bison. A century's worth of work, warfare, and technology replaced 50 million bison with 45.5 million cattle. One wonders what progress is for.

7

Annihilation

No one who has studied western history can cling to the belief that the Nazis invented genocide. —Wallace Stegner, in *Wolf Willow*

Civilized man wrought more destruction on grassland than upon any other natural realm on the continent.

—Tom McHugh, in *The Time of the Buffalo*

We have left less than one-tenth of one-percent of our prairie. The rest of it died to make Iowa safe for soybeans.

—Loren Lown, a man who restores prairie

By 1845 everybody traveling west of the Mississippi understood that the new country was rich in five primary resources—land, minerals, furs, timber and government money—and that of these, the last was by far the most abundant. —Lewis H. Lapham, in *Harper's*

Far from being a child of nature, the West was actually given birth by modern technology and bears all the scars of that fierce gestation, like a baby born of an addict.

We . . . can put realistic names to the social conditions emerging in this West: hierarchy, concentration of wealth and power, rule by expertise, dependency on government and bureaucracy. The American deserts could be made to grow some crops all right, but among them would be the crop of oligarchy. . . . Quite simply, the domination of nature in the water empire must lead to the domination of some people by others.

—Donald Worster, in *The Wealth of Nature*

Recurrence of drought is always dreaded, yet unless one works against nature's plan one need not despair. Extended periods of drought are a part of the plains climate. The grasslands have survived throughout the ages. Slowly but surely the depleted vegetation was always restored. Enough of each species remains somewhere over the wind-swept land to furnish seed for reestablishment of the cover over a period of time. It is only when man aids in the destruction by overgrazing and trampling and by plowing that conditions are worsened and the vegetation is destroyed.

—J. E. Weaver and F. W. Albertson, in *Grasslands of the Great Plains*

I think the prairies will die without grass finding a voice. Its democracy may be against it.

—William Quayle, in *The Prairie and the Sea*

Dependence, hierarchy, genocide, democracy—what have these matters to do with the life and the roots of plants? Jeffersonians and the propagators of the yeoman myth imagined a democracy that rested on a successful occupation of the land. Their assumption was that real political power could flow from the power of a place to the benefit of its inhabitants. They failed to imagine that ignorance of natural community could strip the land of its power, and in doing so cede the power of the inhabitants to others.

A square-yard chunk of big bluestem sod contains twenty-five miles of rootlets, root hairs, and roots. In this simple fact, one may read the flow of power, just as one may read the flow of power in the miles of telephone lines in Washington, D.C. Big bluestem is a grass, not the only one, but the most prominent member of the tall-grass prairie. It is a capstone species, a sort of taxonomical shorthand for what is a diverse coalition of plant life. In the tall-grass prairie, the big grasses include a handful of species on a normal plot, shoulder-high stalks capped by bursts of seed heads. These grasses make up about 90 percent of the living mass of the place, but are really a sort of matrix. These few species of grasses are surrounded by perhaps two hundred species of broad-leafed flowering plants, the forbs, shrubs, and sedges.

All this comes to life in a succession each year, beginning in spring with the short forbs, greening, blooming, and dying before the taller members of the community send them to the shade. Each summer the prairie rebuilds itself from the ground up, always advertising its work with a thick spray of blooms, the brisk yellows of the daisylike composites, the regal reds and purples of clovers, lavenders of the asters, compass plant, leadplant, and gentian. Each plant has its own niche and each depends on the others in some way. Beneath all this is the roots.

The record of this society's centuries is the soil, at once the creation and creator. Soil is the manifestation of the sun's power on the earth. It is our literal link to the only real power of our planet. Through the solar-driven reaction of photosynthesis, plants use minerals provided by their roots, water from wherever they can get it, and atmospheric elements like nitrogen and carbon, mostly to form carbohydrates, which is energy, which is power, the driving force of all animal bodies, including ours. Sun-built carbohydrates raise up the whole system, a system that in turn also falls and dies. Aided especially by fire, but also by a vast array of animals and microbes, the stored energy is broken down and returned to the system as nutrients deposited in the soil. This is where soil gets its organic content and its ability to hold water, which it stores in spaces created by the hollowing of once-living cells.

Because the sun delivers surplus energy, this cycle is not closed, not a balanced equation. Each year the plant community creates more stored nutrients than it uses, so the surplus builds as soil, which in turn enhances the whole system's ability to survive, to live off the savings it has stored away. The lives of past generations enrich those of the present. The force that binds all this to the place is roots, the twenty-five miles of roots and rootlets in each square yard of pure prairie sod.

The basis for many prairie soils is loess, from a German word that means simply "loose." It is a powdery dirt pushed through the plains by the fierce winds during glaciation, then splayed out like flat dunes or a beach. The glaciers left, and the grass made the dust into soil, holding it from the wind with its roots to consolidate the mass and annually investing in it the energy of the prairie's attendant life. Grasslands are soil builders to a degree unimaginable by the first plowmen to invade the plains. Never before had they encountered coal-black topsoil that wrote its record to depths of a dozen or more feet. There *was* a literature of the plains, and this was its library.

In Iowa today there are prairie relicts, not many, but a few the locals call "postage-stamp prairies." Some stand as islands in an unbroken sea of farmed land, but a better simile would be standing as pedestals carved by plows. From some of these relicts, all once level with the surrounding flat land, one must step down three feet or so to walk from prairie onto plowed land.

Settlement was a war on roots. At first the prairie proved implacable, and the yeomen were content to work the fingers of timbers along its edges. Their early cast-iron plows, the lineage in which Thomas Jefferson held a patent, would snap or clog with prairie mulch when set against the fortified roots. The old, heavy cast-iron plows required large teams of oxen and crews of men to run them against prairie soil. Settlers could expect to pay these crews six hundred dollars to clear an acre of tall grass, an astronomical sum then. It was cheaper to clear and farm the forest, a society not nearly so invested in roots.

The new and lighter steel plows of the mid-nineteenth century, however, changed the balance of power in favor of the settlers. They could succeed where cast-iron plows had failed. Early accounts say the sound of the prairie being plowed—there was no roar yet of internal combustion connected with farming—was that of a fusillade of pistols, the pistol-shot cracks of roots breaking.

If the goal of settlement was to bring land under the plow, it was an unqualified success. If we are to hold the myth to its promise, however, there were broader goals: establishing independent human communities and establishing agriculture. Agriculture implies culture, which in turn implies a lasting and symbiotic relationship with the land. Successful agriculture is the integration of the human community into the existing society of soil. In this regard, settlement of the grasslands was and is a failure. The case concerning humans is obvious and quick, but the real problem, the exploitative agriculture that has mined the life from the plains, is more complicated and more important.

The notion that the plains were a great expansion chamber that would contain immigration and expansion of American democracy is false. Immigration, in fact, swelled the existing American cities along the Eastern seaboard, not the interior that Jefferson had so carefully reserved for settlement. During the nation's peak immigration decade, 1900–10, the total farm population from both immigrant and native settlers nationwide swelled by 2.2 million; at the same time, the urban population increased by almost 12 million. The cities absorbed most of the 9.3 million immigrants during that decade.

One study of homesteading in Montana observed that seventy thousand to eighty thousand people flooded into the state to settle on farms between 1909 and 1918, the years when World War I's bonanza wheat prices lured people to the wheat land. Before 1922, sixty thousand of those homesteaders would leave. That is to say, a decade's worth of the great rush to settlement in an area almost the size of California produced a net gain of about ten thousand yeomen.

Similar stories have prevailed throughout the region but especially in the more arid lands, west of the 100th meridian, the wheat lands. Total farm population nationwide peaked in 1916 and has been in decline ever since. Much of the grassland's rural areas peaked in population around the turn of the century or shortly thereafter. Those that hadn't did so by the Great Depression. From that time on towns have been dying, and the rural West has undergone a steady, unbroken decline in population.

In 1893, Frederick Jackson Turner laid down his now famous formula of the frontier, the sort of red flag before the charging bull that was America's industrial progressivism. What molded the national character, he said, was the frontier and our national destiny to fill it. He defined the frontier as any place with a population of fewer than four persons per square mile.

Nearly a century later, in 1980, Rutgers University geographer Frank Popper took a new look at the region that Turner considered and found that 143 rural counties still fit Turner's numerical definition of "frontier," although most of them had been built to a denser settlement and then had become depopulated. One American of every 396 live in these counties, although they cover 949,500 square miles, about a third of the land area of the lower forty-eight states. The area that was to become the paradise of the agrarian democracy now holds a quarter of one percent of its population.

Popper's 143 counties come from throughout the mid-section, but 110 of them form a more or less contiguous band in the middle of the plains that includes the western Dakotas, western Nebraska, eastern Montana, large parts of Kansas, Oklahoma, and Texas, and smaller parts of Colorado, New Mexico, and Wyoming. These counties are the heart of a larger region, the plains, a region that includes a population of 6.5 million people, but these people live in cities clustered around the fringes of the plains: Billings, Cheyenne, Denver, Wichita, Oklahoma City, Bismarck, and Rapid City. That is to say, the grassland West is not predominantly rural, if one is to judge its character by where its people live and are moving.

The grassland is urban; that part of it not urban—by land area, most of it—also is not rural but simply unsettled. Estimates of presettlement population of the region range between one and three people per square mile, about what most of it has now. One must summon some respect for a landscape that is able to dictate its population despite more than two centuries of human efforts to the contrary.

While the people were leaving the grasslands, their plows and their corn and wheat were moving in. During the same period that saw sixty thousand settlers go bust in Montana, the land under wheat rose from 250,000 acres in 1909 to 3.5 million acres in 1919. A war was on. Newly invented steam tractors, whole gangs of them, pulled plows. New varieties of wheat had been imported from the plains of Russia. The decimation of Europe's infrastructure produced an unprecedented demand. From all this, industrial agriculture arose.

Those of us descended from Europeans derive from a wheat culture, a lineage that traces itself in coincident paths with our philosophy, our religion, our law, and our writing. Three domesticated native grasses became the foundations of three cultural traditions in the world. Corn was the foundation of the New World's cities in Mexico and on south to Central America. Rice built the civilizations of the Orient. Wheat built Western civilization.

There are a couple of strains of aboriginal wheat, one first domesticated in Turkey about eight thousand years ago and a second strain farther north, between Armenia and the Balkans, somewhat later. Some would argue that the writing that arose to record wheat transactions in this very region was the creator of Western civilization, but it seems more plausible to lay this to the wheat itself.

Europeans brought their wheat into the American plains, some even by circuitous routes. For instance, religious sects persecuted in Germany were recruited to Russia by Catherine the Great, herself a German. She needed settlers for the forbidding plain in southern Russia, newly won in a war with the Turks. They adapted to the grassland life and developed some new varieties and techniques for farming wheat. Around 1870, however, Czar Alexander took away the benefits Catherine had extended to these people. Recruiters for two American railroads—the Burlington-Missouri and the Santa Fe—heard of this and brought the Mennonites, Amish, and Hutterites to farm railroad lands in the New World.

During World War I, the Turks, the originators of wheat, cut off shipments out of Russia, which was then the world's leading exporter of grain. Europe then turned to America's plains.

Wheat is a short-grass crop, the domestic analog of the mid- and short-grass prairie. It is therefore a plant adapted to the arid reaches beyond the 100th meridian, a place that had already been prepared for the blow. A whole legacy—the Homestead Act, the railroad grants, the industrial revolution, and a war in Europe—sent the plows west out of the corn belt to assault the roots that held together the even thinner and even more tenuous soils of the plains. Thus began what was known as "the Great Plow-up."

By the end of World War I, the nation was farming 74 million acres of wheat, with an annual yield 38 percent greater than only a decade before. More than a third of the total harvest was exported. Some of this increase was the result of the use of new varieties of wheat from Siberia, but most was the work of the plow, new plows pulled first by steam and then by gasoline-powered tractors. In 1914–19, Kansas, Colorado, Nebraska, Oklahoma, and Texas expanded their wheat lands by 13.5 million acres, or about 21,000 square miles. This was not an expansion of the family farm, but further growth of an industrialized farm economy that quickly layered itself into the political geology of the plains. Writes the historian Donald Worster:

> The Great Plow-up, initially provoked by the wartime mobilization of the national economy, might have been expected to pass with victory. Such was not to be the case. The war integrated the plains farmers more thoroughly than ever into the national economy—into its network of banks, railroads, mills, implement manufacturers, energy companies—and, moreover, integrated them into an international market system. When the war was over, none of this integration loosened; on the contrary, plains farmers in the 1920s found themselves more enmeshed than ever, as they competed fiercely with each other to pay off their loans and keep intact what they had achieved. By the mid-twenties, that integration did begin to pay off; having squeezed through the postwar depression, many plains farmers began to rake in substantial fortunes.

The lever for these fortunes was industrialization of the grasslands, and farmers became increasingly dependent on capital, manufactured goods, fossil fuels, and above all, industrialized transportation that, like the refrigerated cattle ships and train loads of bison hides of a generation earlier, could drain the life from the plains. This was not about the creation of local economy and community; no community could eat that much wheat. This was about remaking the grasslands into a wheat factory tied by the railroads to the worldwide industrial economy. It was about mining away the accumulated savings of the grassland that was the soil.

The plow-ups did not abate, but accelerated. Farmers set out deliberately to emulate the ideas of Henry Ford and bring the assembly line to the prairie. For instance, Worster cites the case of the movie mogul Hickman Price, who created a factory farm of fifty-four square miles in Texas and harvested it with a flotilla of twenty-five combines. These enterprises made money and gave rise to the "suitcase farmer," townspeople who kept their regular jobs, spent a couple of weeks planting a crop of wheat on cleared prairie land, and then went back to town to wait for fall and good prices. Between 1925 and 1930, farmers tore up another 5.3 million acres of land in the southern plains. This wild rush to plow continued right up to 1935, when the rains failed and the first searing winds blew.

The dust bowl era was an explosion that many label the greatest environmental catastrophe to befall the nation, at least within our time, and there is plenty of basis for this argument, but it depends on our definition of catastrophe. Indeed it was an enormous upheaval, but to blame it, as many do, on drought is to blame the spark and not the fuel. Drought is as normal to the plains as floods are to the Mississippi and fire is to California's chaparral. Periodic drought is not an aberration but the norm of arid land. Annual mean rainfall statistics are abstractions almost never realized. The rain comes some years in torrents, and some years not at all. The norm is the abnormal. Before and since the dust bowl era, almost every thirty years the whole region has been plunged into sustained drought. The spark was and is always there. What the region had never before known was the fuel: the plowmen's war on roots, which had left the loose prairie soil naked in the teeth of the winds. Agriculture, not drought, was the disaster.

Certainly the cattlemen held some blame for what was to follow. By then the region had been overgrazed for three quarters of a century, with an extermination of many of the native forbs of the prairie and a truncation of the roots systems of the grasses. In earlier droughts, these roots had been the refuge of the plants, where they had retreated to weather out the storm, but the cows had cut off retreat. Short grasses "are short only above ground," write F. W. Albertson and J. E. Weaver in their classic study of the recovery of the plains after the dust bowl era. Their analysis of plots throughout the region did show heavy losses of plant cover and erosion on heavily grazed grasslands. They found almost no damage, even during the dust bowl period, on the ungrazed sections. By far, though, the winds punished the most those lands that had fallen under the Great Plow-up.

During the worst of the dust storms, in March and April of 1935, visibility in the entire mid-section of the nation was near zero. Traffic stopped on the roads. Residents awoke in their sealed houses to find a quarter of an inch of dust on parlor floors. The heavier particles not suspended were swept along the surface of the earth, and the abrasion cut off wood fence posts. From March 20 to 22, the Department of Geology at the University of Wichita weighed the atmosphere above the city and found five million tons of dust suspended over thirty square miles of the city, 170,000 tons per square mile. The dust extended to an altitude of 12,000 feet. The source of the dust was 250 miles west.

It is by now legend in environmental history that this dust did not actually land but swept as a cloud into Washington, D.C., just as Congress was considering the creation of the Soil Conservation Service. One wonders how the history of all of the nation's environmental legislation might be different if the problems would so obligingly present themselves as lobbyists. In this case, the trick worked and the nation began to rethink its ways. There arose after the dust bowl storms first a consensus that this greatest of the nation's natural catastrophes was a direct result of the plow, and then official policy, pursued with particular vigor by the Department of Agriculture, urging farmers to reseed plowed up and marginal lands to grass.

In 1948, the USDA issued its *Yearbook of Agriculture* and titled it *Grass*. The USDA's P. V. Cardon wrote the introductory essay, "A Permanent Agriculture." In it, he said:

> So in the wake of war and in the glow of our unprecedented production, this country looks to the future and considers again the land and its management—this time, as never before, in terms of grass. For around grass, farmers can organize general crop production so as to promote efficient practices that lead to a permanency in agriculture.

After a thorough inventory of the damage created by agriculture on the grasslands, Albertson and Weaver came down firmly on the side of barbarianism's grazers over civilization's plows. They wrote:

> Such conditions induce one to consider thoughtfully the wisdom of the conclusion reached by [the USDA's C. E.] Chilcott in 1927. The Great Plains area has been and should continue to be devoted to stock raising, and all agencies interested in agricultural, social and economic development of this vast region of more than 450,000 square miles should unite in bringing about conditions that will make possible the fullest development of its natural resources for stock production. Crop production should be aimed to supplement livestock production rather than compete with it.

The same writers also noted that in the years immediately preceding the dust bowl era, the short grasses had eerily predicted the coming disaster. Some had curled their leaves. Some sent out tillers, but most simply bunched and placed more room between themselves and their neighbors. This spacing, to these early ecologists, was clear evidence of a peculiar adaptation of arid species. Albertson and Weaver wrote:

> Such an ecological process parallels in a general way that of the human population. Extensive abandonment of ranches and emigrations of settlers had been followed by wider spacings and larger holdings of the remaining population, most of whom had learned by long experience how to endure the hardships of drought.

The dust bowl era spawned a generation's worth of humility and vigorous support of soil conservation by the USDA, but that conservation also worked against the newly confident postwar industrialism. The economic pressures of the postwar boom, like those after World War I, pushed toward exploitation of the land.

Official policy toward conservation notwithstanding, the USDA's L. C. Hurtt would write in 1950:

> Drought, overstocking, uneven distribution, and unwise breaking of native sod are important causes of material deterioration on an estimated 25 million to 40 million acres of range land. Evidence of this subnormal condition is widespread in the form of accelerated soil movement; reduced height growth, or density; change in composition from valuable to less desirable species; and excessive erosion and runoff.

In 1968 the nation began a metamorphosis with the election of Richard Nixon. This period reworked the entire country's definition of itself, but the most relevant aspect of this to the grassland was Nixon's appointment of Secretary of Agriculture Earl Butz. He urged farmers to borrow all the money they could to industrialize their farms and plow them "fencerow to fencerow." That is, he told them to plow down the very grass the government had urged them to plant only a few decades before. And that's what happened. Butz's statement reversed thirty years of federal policy and wiped out any institutional memory of the dust bowl era's lessons. But of course this latest revolution on the plains was not the result of a single statement. It was simply the reemergence of forces that had been percolating since settlement. The Great Plow-up never stopped and the forces that made it never went away. If anything, those forces grew bolder in the postwar boom and Butz was only acknowledging that fact.

The plows and the tractors, however, had become even larger. The twenty-five years since Butz have brought a quantum leap in the scale of farming. Farms have became factories. At the urging of the land grant colleges, extension agents, especially the bankers, seed salesmen, and pesticide salesmen, farmers began specializing in single crops to take advantage of economies of scale. Tractors doubled and then quadrupled in size and price in the space of a few years, as did the array of specialized machinery for dealing with individual crops. This allowed single operators to cover vast acreages in a day. Because of the capital investment required, however, the evolving system discouraged diversification. A given farmer would not raise corn, wheat, flax, beans, and hay when he would need to invest in the special machines and skills to handle all those crops. The single-crop farm became the rule.

Meanwhile, the purchase of these enormous tractors required enormous debt, but drawn by the promise of heavy yields, the farmers borrowed. The surpluses followed, as did agriculture's growing dependence on oil and pesticides. Farmers began spending two dollars in fuel, chemicals, and debt to raise a bushel of corn that would sell for $2.20. In a good year. They incurred enormous costs to work a tight margin, making the whole system pay by volume. When the bad years hit, they precipitated what played out in American headlines as the farm crisis. Bad weather and bad prices could easily make a bad joke of the tight margin, and they did. Farmers simply could not handle the enormous debt and went broke, losing their land in the process. This is the event that sent movie stars to Congress to testify on the plight of the family farmer and rock stars to stages to do benefit concerts. In a single year, 1986, the foreclosure rate on farm mortgages rose to 26 percent of the total mortgages.

Viewed in the short run, this was a watershed event in rural America, changing forever the shape of the landscape, and infusing what remained of the towns with what we think of as city-bred problems like alcoholism, suicide, and crime. In the long run, however, the farm crisis was but a spike in the trend since settlement. It left the nation with still fewer farmers and still bigger farms. Between 1970 and 1987, the year of the last comprehensive farm census, the number of farms in the nation declined by 866,241. For the long term, however, the decline was in synch. Between 1910 and 1987, the number of farms in the nation declined by 4.3 million. There had been 6.4 million farms in the nation just after the turn of the century. There were 2.1 million in 1987.

Nationwide, there has been a slight decrease in land under cultivation during this century. However, most of that has occurred in the East, where farms have been reforested or gobbled by urban sprawl, where soils are heavier, moisture is more predictable, and winds are gentler, where erosion is less a threat. In the plains and in the Midwest, land under the plow has increased since the dust bowl era. As a nation, we farm our most capricious landscape. We have recanted our respect for grass.

The broad numbers of the farm crisis, however, hide the fact that the bankruptcies were more than an economic blip or a weeding of the inefficient. The crisis is not in the banks but on the land, and there it remains.

Let's return to the large tractors. To operate them efficiently, farmers require enormous contiguous fields unbroken by fencerows, ditch banks, and windbreaks. In the seventies and at Butz's urging, farmers tore out these impediments that had been built at the government's urging (and subsidy) in the forties. They leveled low spots and filled brushy draws. There had been strips of brush and grass that formed a bit of habitat, a surviving vestige of nature running the fencerows. All of these irregularities in the landscape had once served as mini-dams of wind and water to break the flow of erosion.

One study in Illinois, for instance, showed that during the period of 1960 to 1987, the state lost about half of its remaining grasslands. These, however, are what are called "secondary grasslands," meaning planted pasture and hayfields. The primary grassland, or native tall-grass prairie that once blanketed the state, was lost during the first wave of settlement. This new loss, coupled with the banishment of fencerows and windbreaks, was seen almost immediately in the precipitous decline in populations of grassland birds. The Fish and Wildlife Service reports that during this period, populations of bobolink fell 90 percent, grasshopper sparrows dropped 56 percent, field sparrows 542.6 percent, and savannah sparrows 58.9 percent. Henslow's sparrows, the once plentiful species of the grass, were too few to count.

Annihilation of prairie species accelerated at about the same time the nation was celebrating its environmental consciousness with the first Earth Day.

Adding to this trend was the pattern of land ownership that developed toward the end of the farm crisis. Credit companies, insurance companies, the federal government, and corporations that provided the capital to the bankrupt farmers wound up foreclosing and owning the farms. The amount of farmland in the nation owned and operated by insurance companies increased tenfold between 1980 and 1986. The USDA reported that this ownership meant the companies employed "in-house management services or outside specialists to get the most return on unwanted assets." Getting this return meant abandoning costly conservation practices such as measures to control soil erosion.

The growing trend of specialization during this period meant that farmers forgot the time-honored tradition of crop rotation, signaling the cleanest break from respect for natural systems and the movement toward industrial agriculture. To understand the significance of this, of all of this and all that follows, we must back up to understand exactly how agriculture relates to natural systems.

Following natural catastrophes such as fire, landslides, glaciation, or volcano eruption, nature begins again. The process starts from what appears to be zero, from a biological desert, and rebuilds with a tool called succession. The polar opposite of this zero point is a system at climax, a dynamic equilibrium characterized by relatively stable and diverse communities of slow-growing plants. This is the situation, more or less, of the tall-grass prairie, with its complement of two hundred or so species, often as many as eight species of plants in one half of a square foot of ground. In reality, the notion of stability here is a bit of stretch; the grassland is always shifting in response to nudges from drought, fire, and wind. Nonetheless, the tools to weather these changes come from its diversity and maturity. Climax is an abstraction, but a useful one in understanding its polar opposite.

Nature's method for dealing with highly disturbed systems is an array of plants called seral species. In the early stages, the bare ground is colonized by a few wind-blown seeds from specialized plants, many of them grasses. These seeds are long-lived and mobile, like shock troops. They evolved to prosper not in communities but in systems of low diversity. They are pioneers. Characterized by very rapid growth, flowering, and seed production, these plants concentrate much of their energy in the seeds needed for the next wave of invasion. Slowly they shade the soil, die, and add their organic content to the developing mat. Succession proceeds on the basis of their bodies. The seral species are replaced by slower-growing, stable perennials that concentrate their energy not on seeds, but on roots.

Virtually all of agriculture is an attempt to artificially prolong this first or immature stage of succession. The grasses we have domesticated are seral species that grow well only in monoculture. They grow quickly and concentrate energy on producing seed. They store carbohydrates in these seeds, which is precisely why we value them as food. From an ecological sense, then, agriculture is a sustained catastrophe. It is the practice of plowing, then preventing nature from healing itself. It is imposition of a monoculture on a system that wants nothing so much as to diversify and stabilize.

Crop rotation is a means to circumvent the worst aspects of this. That is, a given plant—say, corn—has specific requirements of and contributions to the soil. Corn depletes nitrogen. In its living and dying it prepares the community for a new stage. Traditional agriculture would follow a planting of corn with a legume, a member of the pea family, which has the unique ability to pull free nitrogen from the air and deposit it in the soil. That in turn would be followed by a crop of alfalfa and grass to build organic matter.

Crop rotation on the plains is now as antique as the steam tractor. Farmers no longer can afford it; they cannot afford to buy a

$100,000 corn harvester only to use it one year in three. Instead, they maintain early succession with chemicals and traction, both of which are based in hydrocarbons. The chemical fertilizer of choice is anhydrous ammonia, which adds nitrogen to the soil.

Urged on by the petroleum companies that sell anhydrous ammonia and by their own need for higher yield, farmers have begun applying ever higher concentrations of fertilizer, about twice as much as grain crops actually use, according to one study. Because of this, nitrogen has begun accumulating in the soil, but only so much can be held there. The rest leaches as nitrates into aquifers. Nitrates in well water kill infant humans. A survey in 1977 in Merick County, Nebraska, found 70 percent of the wells contaminated with nitrates; statewide in Kansas, 28 percent of the wells showed nitrate contamination in 1986. Nationwide, 20 percent of all wells show nitrate contamination, all from fertilizers.

Succession means other plants wish to break the hold of the monoculture, plants we call weeds. Farmers deal with this through increased cultivation, which in turn exposes more soil to erosion, and more and more, through chemical herbicides.

Monocultures also create a concentrated source of food not only for humans, but insects as well. Insects that had to wander a diverse landscape of hundreds of species to find their plant of choice were kept in check by diversity and by predators such as birds that inhabited the whole and healthy systems. Insects in a monoculture face no such obstacles. Farmers apply pesticides; insects rapidly evolve new generations with resistance or immunity to the pesticides, and the war escalates.

Long-lived organic chemicals that make the pesticides are toxic to humans and have also leached into wells. Most of them are formulated from a class of chemicals called chlorinated hydrocarbons or chloro-organics. These do not exist in nature, so animal bodies cannot metabolize or otherwise break them down. The Iowa Geological Survey found elevated pesticide levels in 39 percent of the five hundred wells it sampled. In the farm belt, repeated studies have shown elevated human cancer levels directly related to agriculture.

Hidden in all of this is the notion of energy. Because many of the chemicals and most of the fertilizer derive from hydrocarbons, modern farming has become not so much a culture as a pipeline for oil. Put another way, the American food dollar is not so much income to farmers as it is to oil companies. Since the turn of the century, there has been an elevenfold increase in the amount of energy used to raise an acre of corn. Add to this our current systems of processing and distribution. The average American eats highly processed food grown in the grassland, well away from centers of population. Farming in the East near the population centers has declined. When the energy to transport and process this food is added to the farmers' energy budget, each calorie of food produced requires about ten calories of hydrocarbon energy.

"Primitive" cultures, in those few places that practice traditional agriculture, exactly reverse that ratio, producing ten calories of food with one calorie of energy. This is a matter of power, literally and politically. A calorie is a unit of energy but with it flows the power to control and exploit. If each part of our food budget is nine parts oil and one part farmer, then who will become powerless in such a system?

And what of water, a form of power that has been organizing society into hierarchies since long before the Organization of Petroleum Exporting Countries reared its head?

In the band of grassland states immediately west of the Mississippi, an excess of water is frequently the problem, as the flooding Mississippi and Missouri rivers demonstrated all over Iowa's cornfields during the summer of 1993. This has produced a sort of drainage society in which farmers busily drain wetlands and swamps to clear the way for corn and soybeans. This act extracts a considerable price from the wetland species of flora and fauna, but is relatively minor in its social implications compared to what goes on farther west, west of the 100th meridian in the plains, the great American desert.

In the drylands, the early American settlers were able to grow wheat without irrigation, at least in some years. The same farmers, however, soon enough learned about the drought cycles that drove them bankrupt in waves, cycles that our public policy began to understand as natural disaster but the land understood as the normal course of events. Meanwhile, the cattle industry had begun to raise alfalfa crops, a wetland crop that required the consistent rainfall that the arid reaches of the grassland were unable to supply. Pervasive still was the national myth that viewed our manifest destiny as the directive to make the desert bloom and people it with democratic yeomen. To do this, we entered into the most undemocratic of bargains: irrigated agriculture.

Irrigation has existed on this continent almost as long as people have. In the deserts of the Southwest and farther south into Mexico, irrigation systems, some of them quite elaborate, organized the corn, amaranth, and bean cultures of societies as widespread and sophisticated as the Aztec and as simple and direct as the Hohokam of what is now Arizona. The latter case, however, is instructive here for what it failed to do: to order the society into layers and set the many to serving the few. Although some canal systems existed in some areas, the flood irrigation systems of many of the small tribes of the Sonoran desert were not so much geared to building canals and ditches as to capitalizing on the natural flow of events. The Hohokam irrigated then, as they still do, by taking advantage of infrequent storms and damming washes and gullies.

The water flowed to a wide range of cultivated native desert plants adapted closely to the place. These would bloom and set seed rapidly after a seasonal rain, provide the food needed, then drop back into the dormancy of the desert. We may think of this as the people's invoking successfully the power of the land, so this power did not accrue to cities and warlords and priests who controlled the flow of water in canals.

It is the later, more elaborate styles of irrigation that built what the historian Karl Wittfogel calls "hydraulic societies." Wittfogel, a German and Marxist before fleeing his homeland in World War II and becoming a rigid anticommunist in the United States, was in his early studies altogether unsatisfied with Marxist explanations of power. He believed Marx failed to account fully for Oriental civilization. After studying China, Wittfogel hit on irrigation and its unique and necessary ability to centralize power and create bureaucracies as a way of explaining far more of China's history than the dialectic. The historian Donald Worster, who teaches in Kansas, has been primarily responsible for applying Wittfogel's analysis to the American West. It is a hydraulic society, centralized and bureaucratized, integrated in a hierarchy of power.

Forces that governed the path of settlement dictated this. When John Wesley Powell considered his plan for settlement of the West, he was heavily guided by his understanding of aridity and successful Mormon irrigation in Utah. It is why he organized his plan for the West around the flow of rivers. But his plan was killed, the West was squared, and the farmers would not go to the rivers. The rivers would be sent to farmers.

This blank-slate notion of the landscape was replicated in a blank-slate notion of horticulture. To this day, a Mexican peasant working in those few areas where traditional agriculture survives uses on the order of forty varieties of corn seeds. So too did traditional farmers in the American grasslands. The farmer's cultural information is attuned to this broad band of genetic information. That is, he understands the landscape, its rolls and tucks, its wet spots and dry, good years and bad, its time of rains and of drought. This web of information dictates his use of seeds. One variety he plants on a south slope, where the sun burns hard, one in a moister draw, one in those years when the rains don't come, one when they do. In peasant markets, all of these varieties then appear in a wild bloom of food that are variations on the theme: corn. Some ears are purple, some stubby, some tough, some sweet. Each in turn has its own use as a food, spawning a bloom of recipes. The diversity of the plant community extends into the human community's diet.

This intricacy of culture is only beginning to be understood in recent years, ironically, because the diversity of the highlands of Mexico is about to be lost. The free trade agreement with the United States is bringing industrial agriculture to Mexico, and botanists have now begun cataloging the genetic diversity of its traditional agriculture the way the Smithsonian cataloged Indians and bison a century ago. This array of cultural and genetic information will be wiped out. The traditional farming communities will be depopulated, and the residents sent to the slums of the cities, replaced with irrigation and tractors.

Virtually all of the hard wheat grown in the United States derives from two varieties—Marquis and Turkey—a spanning of the scale of genetic possibilities from A to B. This replaces the diversity of the prairie.

Ecologists now attempting to restore prairie understand that they succeed only to the extent that they replicate what is called "local ecotype." That is, one may plant seeds of a species, say big bluestem grass, in a given place, and it may fail, because varieties of big bluestem vary as sharply from one another as do varieties of corn. To ensure success, it is not sufficient to replant Iowa bluestem with seed from Nebraska. One must gather Iowa bluestem from local seeds within a few miles of the plot. By evolution through the generations, these have achieved an exquisite state of adaptation to all the weather, bugs, disease, soil chemistry, hydrology, microbes—the band of life that is the community.

American industrial agriculture is the exact reverse of adaptation. Instead of counting on vast cultural knowledge and diverse genetic information to match agriculture to conditions, industrial agriculture remakes the conditions. Nowhere is this more evident than in the case of irrigation, an attempt to impose a uniform, predictable, and bankable rainfall on all of America west of the 100th meridian, America's hydraulic society.

Virtually every free-flowing river in this region has been dammed from its headwaters on down. The grid not outlined by roads is outlined by canals, the water highways that deliver this power to the foursquare farms perched on benches miles from the nearest streams. The rush to irrigate all of the West rose coincident with the progressive politics of the thirties and was fueled hard on that particular postwar hubris that said industry could solve all problems.

We have created a federally funded water system that dwarfs the creations of all other hydraulic societies. Virtually all of the arable West is within the purview of irrigated agriculture. The nation has a total of about fifty million acres of irrigated cropland, and although Florida accounts for a small portion of that, the rest is west of the Mississippi. Much of this has been accomplished by two federal agencies, the Bureau of Reclamation and the Army Corps of Engineers, with a multibillion-dollar investment of taxpayer money stretching back through most of this century. In 1977, the bureau celebrated its seventy-fifth anniversary with an accounting of its activities: 9.1 million acres brought under irrigation at 146,000 farms; construction of 322 reservoirs, 345 diversion dams, 14,490 miles of canals, 34,990 miles of laterals, 930 miles of pipelines, 218 miles of tunnels, 15,530 miles of drains, 174 pumping plants, 49 power plants. By the bureau's own accounting, it had invested seven billion dollars on irrigation alone.

The bureau will argue that this is merely an investment paid back to American taxpayers by the family farmers who use the water to make the desert flower. However, the Natural Resources Defense Council examined that proposition for one district of California's Central Valley project and discovered the following: that the federal cost of delivering water to the irrigators in the district was $97 an acre-foot (the amount of water needed to cover an acre of land one foot deep), while the bureau charged the same farmers $7.50 an acre-foot. Given the size of the average farm in the district, this amounted to a federal subsidy of $500,000 per farm per year. The reclamation law, like the Homestead Act, was set up to assist yeomen and so set its upper limit for a single farmer's irrigated lands at 160 acres. Nonetheless, there is not a single 160-acre tract in the district in question. All of the tracts are much larger, some as large as 30,000 acres.

The dominant crop in the district is cotton, a crop already in surplus, so the federal subsidy in effect makes life harder for cotton farmers in the South, where land is not irrigated. This small example replicates a general trend in federally funded irrigation projects. While the amount of cropland has grown steadily in the arid

West, farms east of the Mississippi have gone back to forest. Worster points out that the opening of federal water projects in the West brought under irrigation almost exactly the same number of acres abandoned by farmers during the same period in the humid Southeast, where the same crops could be grown without irrigation. Federal water means we now grow rice in the deserts of California.

Sandoz may have been right that in the beginning settlement was a way of making the wealth of the West flow east. This is, however, an incomplete picture, now that the real power and wealth of the land has been drained. Settlement now is a way of making federal money flow through the land to the oligarchs who make the desert bloom with oil and subsidized water. Worster has argued that irrigation has built its own power center, which reverses the flow of wealth. The hydraulic society is centered in the West—California—and subsidized by the bankrupt farms of the East and the tax dollars of the citizenry.

Not all of irrigated agriculture is the work of the federal government, so the reach of the hydraulic society is difficult to assess, at least in terms of total investment. On the back end, however, its wake is clear. All of California's hydrology is rapidly becoming one big ditch, and although much is made of the state's urbanization, the golf courses and lawns, 85 percent of the state's water goes to agriculture. Powell's Colorado River is dammed from end to end, and all of its water, from the headwaters in Wyoming to the Mexican border, is claimed by irrigators. It now leaves the nation as a trickle. The Snake, the Missouri, the Platte, the North Platte, the Arkansas, the Gila, all of the major drainages hold major projects. There are virtually no undammed rivers in the West. Agriculture, however, has gone beyond claiming the rivers. Irrigation, too, has become a subterranean war.

There can be no argument that the activities on the lands overlying the Ogallala aquifer can continue. That store of ancient freshwater underlies much of the plains, parts of Nebraska, Kansas, Wyoming,

Colorado, Texas, and New Mexico. Through much of settlement and even through the dust bowl era this area dealt the harshest hands to plowmen. Here the drought and winds bit hardest on the plowed lands. Then the crush of industrial agriculture during the 1970s revolutionized the landscape. Farmers began pumping water from the Ogallala aquifer onto their fields to insulate themselves from dust bowl drought.

Between 1954 and 1971, the number of irrigation wells in west Texas more than doubled. In 1959, Nebraska irrigated fewer than a million acres of cropland; by 1977, the state was irrigating seven million acres. By 1975, the farmers in the six states above the Ogallala were pumping more than twenty-seven billion gallons of water a day from it. In places, the water level in the aquifer dropped four feet a year while average rainfall replaced a half inch. This deficit, the annual depletion of the aquifer, equals the flow of all of the Colorado River, but the Ogallala no longer has its headwaters. Laid down at least three million years ago when the climate was wetter, it is largely sealed from the surface waters of the plains. It is being mined.

By 1980, it was watering 20 percent of the nation's irrigated cropland. Hydrologists suggest it will be depleted within thirty years.

It has a pretty name, the Palouse Prairie, but the place is no longer pretty. It is the heartland of a grassland, but there's little grass here anymore. Oddly, it lies west of the Rockies in eastern Washington State and north Idaho, but it was the birth land of a range of grassland species that dominate stretches of the high plains east of the Continental Divide. Blue bunch wheat grass, Idaho fescue—a pretty bronze burst of a bunchgrass—rough fescue: they once were the backbone of a community called Palouse. In various combinations they could be found ranging clear east to the Black Hills and north to the Sweet Grass Hills, tough old bunchgrasses on a mean and unforgiving terrain. In any aspect except rainfall, this is the toughest of the grassland climates, but then blue bunch wheat grass makes it south into the Great Basin of Nevada and Utah, and that's the toughest all around.

In the northern Rockies the soils are thin, wispy, and dry, as though only yesterday they were rock, and some were. The Julys and Augusts as often as not bring no rainfall to speak of, or when they do, it's thunderstorms, lightning, and flash fire. The Decembers and Januaries often as not bring temperatures of forty below, backed by fifty-mile-an-hour winds. In all this the Palouse plant communities still survive, but mostly not in the Palouse proper. They inhabit the tawny fingers of valleys that break the fir and pine slopes of the Rockies.

The Palouse itself has been given over to wheat. It is not a flatland prairie, but a series of humpy little hills, some so steep the farmers work them with steel-tracked bulldozers instead of rubber-tired tractors. Notions like contour farming, shelter belts, and fallow strips—erosion control measures pushed by the government in the wake of the Dirty Thirties—are here forgotten. The farmers plow fencerow to fencerow where there are fences and just plain plow where they are none.

There is a joke among the wheat farmers of the Palouse. When the wind blows, they like to say: "A lot of real estate is changing hands today."

This is as good a place as any to ask a fundamental question. The democratic idea of Jefferson and Crèvecoeur has devolved upon us as mortgages, foreclosures, pesticides, suicides, dust bowls, subsidies, and surplus. The farmer and poet Wendell Berry, in an important essay, "What are People For?" considered all of society: its cities, farms, and the long reach of trends that rendered people obsolete, first on farms, then in cities, as people became the legions of unemployed, criminal, and homeless. The question that framed all this is: If there has evolved a social system that, in order to exist, must render people obsolete, then what are people for? In the Palouse and throughout the grasslands, we ask the same question, reframed: What is agriculture for?

Agriculture implies culture and culture implies learning, yet standing in the Palouse, it is difficult to imagine what we have learned. Annual topsoil loss to wind erosion here in some years totals one hundred tons per acre, or an inch of topsoil removed to the wind every 1.6 years. When the plowmen first broke out the Palouse in the 1920s, the topsoil the grass had built beneath the moldboards was thought to be so deep as to be inexhaustible. Like the Oglala, however, it now approaches exhaustion. Hilltops are beginning to show pale streaks of subsoil. Productivity is declining. Without topsoil, agriculture is impossible.

In 1982, the Soil Conservation Service, the agency born when the grassland's dust blew into Washington, D.C., completed a comprehensive inventory of erosion nationwide and found that the average of all of the nation's croplands was a total loss of eight tons a year per acre to erosion. These are numbers that rival those of the dust bowl years, and one wonders what we have learned.

To help counter this, the USDA worked into the 1985 Farm Bill the Conservation Reserve Program. The 1982 inventory showed that 101 million acres of "highly erodable" land had been brought under the plow in the wake of Earl Butz's famous pronouncement. The program recognized this was not in the national interest and agreed to pay farmers an average of $49 a year per acre of this land they would return to grass, an adept piece of blackmail on the farmer's part. Five years into the program, the government was paying $1.7 billion a year to reverse Butz's legacy.

The notion of culture implies an accounting for all of the forces that support us: the grass communities that made the soil, the soil beneath, the human understanding of these, and the human communities that devolve from all, yet if farming does not create such a lasting relationship with the creation, then what is farming for?

Is it for the farmer? If farming supports farmers, then why has it been so hard on their numbers? If a disturbance in a natural system had caused so precipitous a decline in the number of animals in that system, we would be concerned. If tissue samples from birds were coming up as full of pesticides as are those from farm families, we would be alarmed.

Or maybe this is just a weeding out, a harsh system meant to create a new animal that can survive the system. A friend of mine says there are no stupid farmers left, that you can't be stupid and still farm the government. It takes some intelligence to fill out all the forms.

Direct federal subsidies to farmers—not counting the water projects, the roads, the federally funded research—totaled $11.9 billion in 1988. The American Farmland Trust estimates that the federal government spends $40 million per day on subsidies to yeomen. This in part accounts for the fact that those remaining on the land have a decent standard of living, if such a thing can be measured in cash. As of 1988, the average farmer was making close to $40,000 a year—most of them, contrary to popular perception, with very little labor. About half of that, however, comes from income generated off the farm. A monoculture wheat farmer spends a couple of weeks in spring or fall plowing, fertilizing, and planting and a couple of weeks in the fall harvesting. A friend says a modern Iowa farmer's idea of crop rotation is corn, beans, and Florida. The suitcase farmer still exists.

Given that, can farming be for community? The gross population loss in the grassland states underestimates the impact in farming towns, the villages so central to the yeoman myth. These are even harder hit in that many of the larger cities in the region are growing. From Texas to Montana, the plains are full of towns boarded up and forgotten. Community is now the satellite dish, CNN, the C-Store out by the interstate, and the interstate itself, which takes one on shopping trips 150 miles away.

Income notwithstanding, ask a farmer if he feels secure or in control of his future, and then begin to understand the anger and angst of the heartland. Farming cannot forever be waged as it is today waged. The Oglala, the credit crunch, the erosion, and the pesticide resistance being bred—all will catch up. So will the federal deficit. The yeoman myth has sustained the farm subsidies long past their usefulness.

Farmers do not originate or control the power that derives from the land. That passed with the roots and the grass, the source of soil. Real power in agriculture is a knowledge of the plants, and that power has been systematically stripped from the heartland. The substitute is the alien power of federal water and hydrocarbons, the calories the oil companies provide. Wrapped in this is an understanding among farmers that they serve on the sufferance of capital, oil, and subsidy and could just as easily be beheaded by the prince.

Is agriculture, as Jefferson supposed, for democracy? Not when it has created oligarchy in the form of concentrated wealth on the plains. Average-size farms, those with less than $100,000 in sales, accounted for only 21 percent of the total sales in 1987; megafarms, selling more than $500,000, accounted for 38 percent of agricultural sales. Farming cannot create a Jeffersonian democracy when it has funneled the federal treasury through the hands of farmers and into the hands of suppliers of oil and machinery.

Is agriculture for food? Do we need more food? The grain belt has produced a surplus of food in virtually every year since 1920. That's why the subsidies exist. So why in the face of this do we bring more land under the plow? Certainly there is hunger in the world, but that hunger has existed throughout the time of surplus. American grain has been a prime source of aid for world hunger, but some world aid officials argue it also has been the cause of world hunger: The surplus grain disrupts traditional markets and puts Third World farmers out of business, just as it did American farmers east of the Mississippi. Further, the dispensing of that grain in Third World countries itself creates a power structure that is inherently disruptive of traditional power structures.

American agriculture, with all its technology, subsidies, and labor, supports a population of 45.5 million cattle in the plains states, the same area that held 50 million bison without any of these. In many ways, though, cattle themselves provide a way to hide surplus grain. There is no need to feed grain to grazers. The bison that were the basis of the traditional plains economy ate no corn, nor did the Mexican cattle that grazed the grasslands for three hundred years. The habit of fattening cattle on grain is just that, a habit. As a result, we eat fatter meat and the heart-stopping cholesterol it contains. Seventy percent of the grain crop of American agriculture goes to the livestock that replaced the bison that ate no grain, and one wonders, what is agriculture for?

There is indeed a farm crisis, but it did not begin with the foreclosures and benefit concerts of the 1980s. It began with the invention of the steel laminated moldboard plow.

8

Aliens

I am standing in a Kroger store in Charlottesville, Virginia. On a bit of a hill just outside of town stands Thomas Jefferson's estate, Monticello. Surrounding it are the gardens that were tended by his slaves and by his own deep commitment to agrarianism. To visit these is to understand that his horticulture was his politics, his plants a cultivation of power. From these were to flow straight down the hill the blessings of agriculture replicated in a seamless geometric pattern from here to the Pacific.

The red rich soil of Virginia, with its abundant rains and gentle climate, today holds few farms. Like most of the eastern seaboard, it has returned to second-growth forest that only barely camouflages the suburbs. Jefferson's agriculture, which was to be the basis of all community, has ceded these communities. The nearby Blue Ridge Mountains and the intervening valleys hold second homes, condos, and golf courses. The remaining pastures hold the high-dollar horses of would-be squires in would-be colonial homes—lobbyists, accountants, and bureaucrats only one jump ahead of the suburban tsunami raised by Washington, D.C., to the north-east.

On the loudspeaker in the Kroger store a woman announces a special sale, not just on melons but "California melons," food from the irrigated fields of California.

Jefferson once said: "The greatest service that can be rendered to any country is to add a useful plant to its culture," and in doing so set the seed of his own idea's undoing. A friend who is a botanist says: "A weed is a plant out of place."

President Lincoln created the Department of Agriculture and signed the Homestead Act, the railroad land grants, and the legislation creating the land-grant college system, all within a few months of each other; this phalanx of laws would make him the father of industrial agriculture. By the beginning of the twentieth century, one of Lincoln's contributions to the agrarian legacy—the Department of Agriculture—had evolved to match the Progressive politics spearheaded by that transplanted westerner, Theodore Roosevelt.

By measure of his regard for nature, Theodore Roosevelt was our best president. His commitment to conservation was real and unshakable, and he maintained a connection to nature that has not been equaled in the White House since. The Clinton administration, when faced with ranchers' opposition to grazing law reform in 1993, ducked for the cover of Congress and "consensus building." Roosevelt called up federal troops to arrest ranchers.

Roosevelt's commitment, however, was flush with the hubris of the day, a belief in the perfectibility of humans coincident with the perfection of the landscape. Nature was the big garden that was to be tended, a park to be manicured. The trees that neatly lined streets in front of white picket fences were to flow in rows into forests that were harmonious replications of peaceful communities. From these notions of Roosevelt's time rose such ideas as "sustainability" and "sustainable communities." Roosevelt gave us the nation's first forester, Gifford Pinchot, and the Forest Service, an organization that was to couple the productive capacities of nature to orderly human communities as a motor is coupled to the chassis of a purring Model T.

Roosevelt's Agriculture Department began to understand some of the troubles yeomen suffered in the West. Rain had not followed the plow, and homesteaders had not prospered on the 160-acre plots of Lincoln's Homestead Act. In part, at least, the Agriculture Department proposed to solve these bumps in the road to earthly perfection not by rethinking the agrarian ideal to match the dictates of the West but by remaking the botany of the grassland to match the ideal.

This agenda survives. In his 1990 book, *The Range*, the rancher Sherm Ewing propagates the notion by insisting that not just the West but all of the New World was the "have-not land" lacking in useful plants. Farmers were forced to bring Old World seeds to render the place habitable. He lionizes Timothy Hanson, the proselytizer of the English grass timothy. Ewing even cites potatoes as one of the European settlers' contributions to the New World's floral landscape, though in fact potatoes originated in the New World in South America and were imported to Europe, dramatically affecting the Old World balance of power. Because they were a dense bundle of portable carbohydrates, they greatly increased the mobility of armies and became a sort of botanical secret weapon.

In 1898, the Agriculture Department launched a campaign that was to be the crest of a wave of introduction that has made the nation's mid-section the unwilling host of literally thousands of species of exotic plants—in a real sense, thousands of species of weeds.

An ecosystem arises from a unique set of conditions—climate, soil, heritage, and geography—to raise for itself an equally unique plant community adapted to the specifics of place. Because of this, ecosystems tend to be relatively immune from extensive invasions by plants from neighboring ecosystems. The prairie can't be taken over by maple trees because the prairie won't grow maple trees. By the same token, the nation's grasslands were saved from weeds already being fought by eastern farmers, largely because those weeds were European and so required the moisture of the East.

By the beginning of the twentieth century the USDA realized it no longer could rely on crops from Europe to colonize the arid plains but would instead have to go to a place that closely matched the grassland's conditions. To do this, the agency created a division aptly named the Bureau of Plant Industry.

Frank Meyer was a Dutch immigrant whose lifelong love of plants finally landed him a job with the Agriculture Department, which in turn led to his spending the high points of his professional life in China, Mongolia, and Siberia. By 1905, the year of Meyer's first expedition, it was generally believed that this exotic stretch of the world held the future of the American West. The USDA was turning to the grassland of Asia, recognizing, as the bison and horses who crossed Beringia once did, that it was the closest analog to North America's arid mid-continent band.

Meyer made four expeditions, the first a three-year stay in China beginning in 1905. He was in Europe, Siberia, and China from 1909 to 1912; Siberia, Manchuria, and China from 1912 to 1915; and finally in China from 1916 to 1918. As a result of these four trips, he single-handedly introduced 2,500 species of plants to the United States, a record that would have earned him Jefferson's praises as the nation's greatest servant, but a modern-day botanist called him "Typhoid Mary." Both interpretations are possible.

It can be argued that work such as Meyer's was a significant factor in the Allied victory in World War I. Charles Saunders of the Canadian Department of Agriculture used Russian wheats, some of which Meyer collected, to develop an early ripening variety of wheat known as Marquis. It is one of two cultivars that today are the basis of virtually all wheat farming in America, but Saunders developed it just before the war created the wheat bonanza on the American plains. Without it, American farmers would not have been nearly so successful producing the grain that helped win the war.

Still, much of what lay behind Meyer's work was the blank-slate approach to the American West, that it simply had to be torn down and rebuilt botanically from the roots up. Said Meyer during an early visit to a place in China, "I went to Wu Tai Shan and saw and went away, for it was as barren as the plains of Nebraska."

The area in question happened to have been deforested, and the absence of trees was in Meyer's and the USDA's mind the equivalent of "barren." Much of his work was concentrated on finding varieties and species of trees that would survive the arid plains. In a Peking nursery, he found *Syringa meyeri,* which every plains housewife came to know as a lilac bush, the fixture of farmhouses from Texas to Manitoba. He found the Kashgar elm in Afghanistan. Meyer's biographer Isabel Shipley Cunningham said: "He felt sure the settlers in the American Southwest would appreciate the dense foliage and the striking umbrella shape of this tree." Turkestan gave him the Khotan ash, which the USDA proceeded to plant by the thousands in Nevada. His Siberian elm became the tree of choice for windbreaks throughout the plains states. Following the dust bowl years, the USDA planted 17,000 miles of windbreaks of Chinese and Siberian elms. He brought the Russian olive, a chestnut, hickory, catalpa, lemon, hawthorn, persimmon, ginkgo, the honey locust, the Siberian larch, a privet, poplars, willows, rose bushes, and oaks. Meyer's introductions of bamboo survived as ornamental plantings around Washington, D.C., and became fodder for the famous pandas that would eventually reside at the National Zoo.

One might legitimately ask how there is harm in all of this, something as simple and basic as a farmwife planting lilacs round her doorstep? In some cases, there probably is none, other than continuing our European worship of trees. Those lilacs disconnect one's yard from the prairie that is around and so disconnect our lives from reckoning with the real wonders of the grassland. The Nebraska plain is not barren, after all.

The importation of new plants in most cases probably is simply harmless silliness, but in the cases of a few species it carries the potential of catastrophe. Meyer brought with him botanical bombs that explode even today.

Exotic plants have evolved in their own set of circumstances, and through the generations have learned methods of survival particular to those places. Taken from those circumstances, they often use those skills to dominate their new surroundings. This alien knowledge is a leap in technology, just as stone-tipped spears were a leap in the Ice Age and smallpox was during European settlement. The exotics from the Eurasian grasslands were completely adapted to the arid circumstances of the West, so unlike earlier European invaders, could colonize and reorganize the landscape. Chinese chestnuts evolved an immunity to a blight indigenous to China, but Meyer's specimens brought that same blight with them to the United States. American chestnuts had no such immunities and today the American chestnuts that were the dominant trees of the eastern deciduous forest are gone. Tumbleweed, that essential symbol of the free-rolling West, is an imported Eurasian. Its name is really its strategy. In fall it dries and breaks off at the stem to let the winds roll it and its seeds across the land. Picturesque as it may be, it still is a noxious weed that supplants native vegetation. Tumbleweed is in the same family as mustard. Meyer found mustard during a 1911 trip on a steamer down the Volga River, the same weed that plagues midwestern farms. The context of those plants in Eurasia included enemies that kept them in check. Stripped of that context in North America, they became weeds.

On that same trip Meyer collected crown vetch and crested wheat grass, a fact that can cause most present-day ecologists to wish for the sinking of that Volga steamer. Crown vetch is a legume, a nitrogen-fixing forb that grows readily on disturbed sites. In the plains, people concerned with erosion began planting it along road cuts. Using the network of roads to spread, the vetch took over, and began creeping onto surrounding lands, supplanting all else. It invades native stands of prairie. People seeking to restore prairie must attack it with herbicides.

Crested wheat grass is another matter. Under grazing, even overgrazing, it is an increaser. That's why the USDA began singing its virtues just after Meyer and an earlier USDA collector, Neils Hansen, brought the blessings of this grass to the New World. The Agriculture Department began planting it on abandoned wheat land in North Dakota in 1916. By 1929, commercial seed was available. Douglas Dewey, a modern-day plant geneticist for the USDA, still offers the company line:

> The grasses of the steppes and deserts of Central Asia have developed and flourished under heavy grazing by cattle, horses, sheep, camels and what-have-you for thousands of years; that's the best reason why those species are more suited to our western livestock industry than North American range plants. It's true our own grasses developed under sometimes heavy grazing by wild animals, but I'm sure the pressure of domestic—even nomadic—livestock is much heavier. Domestic animals under any system of management are less mobile than wild game. A cow just looks at our good old native Bluebunch wheatgrass, and it about wilts on the vine, so to speak—can't really stand much grazing. On the other hand, you can grub Asiatic crested wheatgrass right into the ground and it will bounce right back.

Fences, the Jeffersonian grid, and the importation of European livestock were all violations of the prime law of the grassland, the law of mobility. Because the cows had to stay in the small plots of the Homestead Act, they grazed in patterns the grassland had never known. The solution was crested wheat grass.

A generation after this grass's introduction and after the imported wheat had stimulated the Great Plow-up of World War I came the dust bowl years and the abandonment of untenable wheat farms. Instead of returning to native grasses, however, much of this land went back to crested wheat grass, at the USDA's urging and by virtue of its supplying cheap seed. Ewing quotes a fellow rancher's memories of that time in Montana:

In the thirties up here on the bench Pa says, "Damn we gotta do somethin' about that land. We're payin' taxes on it and it ain't producin' nothin'." . . . After that we bought some crested wheatgrass seed and Pa got an old Case tractor and a four-horse drill that held a ton of seed; we put a tongue in it for the tractor and I went to seedin' day after day.

In the 1980s, the Soil Conservation Service came to realize that the Great Plow-up following Earl Butz's advice to plow to the fencerows had brought highly erodable land under the plow, and it began the Conservation Reserve Program. Farmers were paid to take land out of production and seed it to grass, in most cases crested wheat grass. After three quarters of a century of seeding, this exotic is a part of the landscape. Fields of it run virtually in monoculture for miles along the high plains, as if it were a part of the botany. Unlike the native grasses, crested wheat grass will not cure—that is, it will not carry its nutrition into the winter. That was the secret of the natives, the deal they cut with the community that allowed the bison to use their big shovel heads and survive the high plains blizzards. Much of that deal is done, however, on much of the land. Now in some winters the mule deer, elk, and antelope can wander down to winter range from the snowbound hills and stand and starve to death in endless fields of grass.

A study by the Fish and Wildlife Service in 1993 of twelve species of birds endemic to the Great Plains discovered precipitous declines in their numbers over the past twenty-five years. Grassland bird populations were found to be more disturbed than those of any other American biome. The study blamed this decline on the conversion of grasslands to agriculture and the planting of crested wheat grass.

Kansas senator John James Ingalls was clearly on the right path when a spring ride in his home state's grassland prompted a florid speech, first printed in 1872. A section of it has become a sort of grass man's creed and often is quoted as such:

> Grass is the forgiveness of nature—her constant benediction. Fields trampled with battle, saturated with blood, torn with rust of cannon, grow green again with grass, and carnage is forgotten. Streets abandoned by traffic become grass-grown like rural lanes, and are obliterated. Forests decay, harvests perish, flowers vanish, but grass is immortal.

Remarkably, in the same speech Ingalls goes on to express a sort of bioregionalism, suggesting that even philosophy, religion, and the character of a place's inhabitants spring directly from the natural character of the place, and the mediator of this process in America is grass. Further, he asserts a hierarchy of plants and even of grasses—that some are more noble than others—and one can guess where he is headed, even forgive him this small conceit. He is, after all, in Kansas, native home of big bluestem and Indian grass, plants that at the time covered much of the state and rose so tall by August as to hide the back of a standing horse. It wouldn't be the first time—or the last—a senator's speech became a bit hyper-inflated with home-state pride.

Ingalls, however, was not talking about the tall grasses of Kansas at all, although his speech did refer to a specific species of grass. "It was Blue Grass," he said, "unknown in Eden, the final triumph of nature, reserved to compensate her favorite offspring in the new Paradise of Kansas for the loss of the old upon the banks of the Tigris and Euphrates."

The speech was part of the senator's campaign to rip the tall grass off the prairie and seed Kansas to Kentucky bluegrass, that grass that is neither blue nor native to Kentucky. "Blue grass does more than nourish splendid horses. It supports a whole society of muscular men and voluptuous women; upon its foundations rest palaces, temples, peaceful institutions, social order."

Social order.

Bluegrass, *Poa pratensis,* is an increaser. Like crested wheat grass it has a low growth point, meaning new cells are added to leaves at their bottoms, pushing up the whole plant. Decreasers have high growth points; new growth occurs at the tips. Such is the case with most native species, adapting them to infrequent grazing, as occurs with bison. When a cow clips off a bunch of bluegrass, however, it simply sends out new growth, which is exactly why bluegrass is also the basis of most lawns, what the suburbanite knows as grass. It can be mowed to yield the crisp, clean manicure of an English country garden—that is, social order.

Even without benefit of speeches from U.S. senators, bluegrass spread through the plains. Its seed is mobile and easily harvested. Grazing cows would select for it as if hired to weed out the natives. In bags of seed in wagons and rail cars or clinging to the hair around horses' hooves, the seeds of bluegrass worked their way west along the roads. In 1992, a pair of researchers documented the extent of the invasion. High alpine meadows of Glacier National Park were being overtaken by invasions of bluegrass that had followed the horse trails deep into the mountains. Packers had deliberately seeded the grass along the trails so their horses might have something to eat.

Cheatgrass is a monster that is native to Eurasia and today has a wider range in the New World than any native grass. It can be found in all of Canada and the United States, except in the seven states of the extreme Southeast. In areas of the Great Basin it has become the dominant species. Because it thrives on the bare soil of disturbed areas, it follows roadbeds, but it quickly spreads upland. Even cattlemen hate cheatgrass. The awns on its needlelike seeds burrow into the fur and jaws of animals, causing injury and, in some cases, life-threatening infections. Except during green-up in early spring, its nutritional value is virtually zero, especially for wildlife. In some areas of Nevada, Idaho, and Utah, it is the leading threat to wildlife, especially in winter range. Whole herds of mule deer dwindled to almost nothing when cheatgrass invaded winter range. One area I know of in Nevada has been rendered a moonscape by open-pit gold mining, the largest such collection of mines in the world, and still federal land managers in that region consider cheatgrass to be their leading environmental problem.

Probably a stowaway in imported wheat seed, it got its name in the mid-nineteenth century when it began showing up in wheat fields in Washington state. Because it could successfully compete with wheat, it cheated the farmers out of their yield. The only better name for grass I know is that of a weedy species that escaped from an agricultural research station in Utah. Nearby farmers named it "professor grass."

Cheatgrass, also known as downy brome or broncograss or *Bromus tectorum*, serves as a sort of natural revenge on the cattlemen of the Great Basin and Snake River plain in southern Idaho. There is no record of pre-settlement grazing to any significant degree in that region. In contrast to the plains east of the Rockies, its plant communities were not adapted to the upheaval of bison. Because of this, grazing in the nineteenth century decimated the high desert of the Great Basin, and it still does. The problem is particularly acute there because the dominant species before the cows and the cheatgrass came were bunchgrasses. The spaces between the bunchgrass were occupied by a soil and plant structure peculiar to the high desert, a coating of cryptogamic (hidden-seed) lichens. These lichens sealed the soil from invaders but are exceedingly fragile, some too fragile to withstand even the sole of a hiker's boot. Recently the National Park Service found a relict ungrazed grassland in Utah and has kept its location secret so that curious hikers will not upset its balance.

Elsewhere, though, cows' hooves shattered the lichen, the scheme of the place, leaving bare soil between tufts of bunchgrass. Into these bare spots came cheatgrass seeds to pursue the second step of the species' unique strategy. It grows quickly in spring, sets seed, then dies, effectively getting the jump on any native competition. Thus, by late summer it stands dead, a rich wine red all across the sagebrush hills. In late summer the region gets almost daily doses of afternoon thunderstorms, most of them dry but full of lightning. As is the case with most grassland, fire is a way of life for plant communities of the Great Basin, and they are adapted to survive its onslaught, but a given site in the region saw those fires once every sixty to 110 years. Because bunchgrasses and sagebrush are not contiguous, they carry fire poorly. The dry, dead cheatgrass, however, has boosted the fire frequency to every three to five years, and because it is thick and dry, it allows hotter fires. Hotter fires kill more plants and make more disturbance, which in turn makes more space for cheatgrass.

By now, this spiral has screwed its way to a critical infestation of a hundred million acres of land in the region. So far. No one has come up with an effective way of eradicating the stuff, short of plowing and farming the land, a cure worse than the disease. There are certain stop-gap measures that show some results, but these are expensive. For instance, the Bureau of Land Management spent $100,000 in one year trying to beat back cheatgrass from 1,700 acres of the Snake River Birds of Prey area in southern Idaho. That represented 90 percent of the range management budget for a subdivision of BLM land that includes two million acres, 90 percent spent on one tenth of one percent of the land. Most of the two million acres is infested by cheatgrass.

"There's not enough money in the U.S. Treasury to get cheatgrass out of everywhere," Mike Pellant, a BLM range conservationist, told *The Atlantic Monthly*.

Orchard grass is a European native that was under cultivation in Virginia by 1760. In the late nineteenth century, the U.S. Forest Service decided to use it instead of native species to reseed abandoned logging roads in the northern Rockies. Soon after, it was discovered that black bears like orchard grass and could be found bunched up on the old roads during spring, grazing the new growth. When the region's bear hunters learned of this they simply walked the roads and shot the baited bears.

Johnson grass is a wild sorghum, considered a weed through much of the southern grasslands it has invaded. It came originally from Turkey and was spread along railroad beds during construction. It was in the hay the railroad crews used to feed their mules.

Bermuda grass is considered one of the most important grasses in the South. It came from Africa, because it was used as bedding on slave ships.

The biomass of a grassland system—that is, the sheer volume of plant material raised up by a place—is mostly grass. The diversity, however, the number of species, is increased by the forbs, the leafy plants. The grass provides the bulk that feeds the creatures. It is the productivity. The forbs fill the spaces between the grass and provide the biological services. They fix the nitrogen and minerals; they sponsor the fertility of the land. Because of this, systems can adjust to a degree to some invading grasses. Bluegrass, for instance, when not overgrazed, still provides the matrix in which some of the attendant forbs can grow. Some exotic grasses are compatible with native forbs. There are, however, invading forbs, and with some of them the consequences for the system are dire.

Spotted knapweed is a Eurasian, a weedy bush, a relative of the flower gardener's bachelor buttons. It is dotted in late summer with blooms that look like miniature pineapples, topped with a tuft of bright purple fringe. For most of the rest of the year, however, it exists as a sort of dull gray haze lingering over infested fields, of which there are many. Spotted knapweed can be found throughout the country, but true to its Eurasian roots, it thrives in the more arid grasslands of the high plains and Palouse country. It has infested more than seven million acres in Montana and the surrounding states and provinces.

Knapweed, like most invaders, prefers disturbed sites, so the gray haze is a marker of roads, overgrazed fields, logging skid trails, and dirt bike trails. Its seed walks up these like a rat up a ship's ladder, spreading its tough woody taproot to any space that will hold it, and most will. Like most exotics, it has no natural enemies, although some insects have been introduced from Eurasia in hopes of bringing it under control. They do not, because nature does not evolve predators to eradicate their prey. So far, the insects fail to keep it in check. It is, however, relatively easy to kill with herbicides, so it can be controlled on accessible and farmed valley floors.

Elsewhere it seems to go where it pleases, and most grazers won't touch its spiny and woody stems. A researcher in western Montana sampled several typically infested fields of knapweed—and fields in that area are typically infested—to find as much as 1,920 pounds of knapweed per acre. The same infested field produced 120 pounds of grass per acre. A year later, after the knapweed was killed with an herbicide, the field produced 2,330 pounds of grass per acre; the knapweed had suppressed 95 percent of the grass.

Grass is the biological medium of exchange of the grasslands. Cows or elk, it doesn't matter, when the grass falls, the capstone species fall with it.

So far, only two plants seem capable of competing with knapweed in areas hit hardest by it: sulfur cinquefoil and leafy spurge. The Eurasian sulfur cinquefoil made its way into the grassland by establishing itself in the East and spreading through the upper Great Lakes region to the plains. By the 1950s, it had spread as far as British Columbia. Like knapweed, it prefers the West and has settled there. Densely. Some infested areas produce as many as twenty-five plants per square yard. It sets a deep green marijuana-like leaf and in late summer pale yellow flowers. The plants can live as long as forty years and annually produce as many as 1,650 seeds each. Virtually no animal eats it. Very little is understood about its control, other than the fact that heavy doses of herbicide are about all that will kill it. Unlike knapweed, however, it is not necessarily dependent on disturbances such as overgrazed fields or roads as vectors of its transmission. It will invade healthy grassland communities, and these do not remain healthy much longer.

Leafy spurge is also a Eurasian, which now grows on every continent of the world except Australia and Antarctica. Its introduction into the United States was accidental but still a product of our penchant for travel, probably arriving in soil used as ship's ballast. It was discovered in Massachusetts in 1827 and collected as a rare plant in New York in 1876. By the early part of this century, it had reached the grasslands, an environment similar to its Asian home, and there it thrived. It has no enemies. None. It has shown a penchant for proliferation at the exclusion of all else that is practically human.

Ninety percent of the leafy spurge infestation in the United States is found on the upper Great Plains, in a circle 1,200 miles in diameter, with its center at Wolf Point, Montana. Within this circle, it severely infests at least 2.5 million acres of grassland. The definition of "severely" requires an up-close look at a site.

Like most exotic forbs, leafy spurge, *Euphorbia esula,* is particularly adept at occupying disturbed sites, so it follows the roads, but it has a second trick. The seed is bulbous and light, so it will float. It rides the rivers and canals in high water and washes out onto the floodplain. Miles of riparian habitat, the rarest and most critical habitat for the region's wildlife, have been carpeted by spurge. Once it has moved into the grassland, either along roads or irrigation canals, it can establish colonies in healthy, ungrazed stands of grass. Mature plants send down roots as deep as fifteen feet. Spurge competes with and shades out all other grasses and forbs. Severely infested fields raise only spurge and a brown carpet of moss that survives around the stems.

If pulled, as in grazing, the stem breaks, leaving the deep taproot. The wound secretes auxin, a hormone that causes not one but several new stems to grow. Spurge is, however, seldom pulled by grazing; only some species of domestic goats will eat it. Because it contains latex, the weed is an irritant or even a poison for cattle and humans. Infestations reduce the carrying capacity of livestock range by 50–75 percent. Even splotches of grass that remain among the spurge go ungrazed, because cattle tend to leave an area where it is growing.

The sole means of controlling the weed is heavy applications of the herbicide picloram. Picloram, which Dow Chemical markets under the trade name Tordon, cannot be used in floodplains, nor can it be practically used in some of the rugged and remote areas of the West in which spurge has spread. Tractor-mounted sprayers can't climb the hills. Cropdusters can't fly the steep canyons or apply the herbicide with the precision that responsible use requires.

Ranchers in Montana, North Dakota, and Wyoming spend more than nine million dollars a year on chemical control and still suffer estimated annual forage losses of about that amount. And still the spurge spreads. Into national parks, into wilderness areas, and into biological preserves. It does not simply disrupt the delicate balance of these ecosystems. It reduces the places it takes over to biological deserts.

There is a growing school of science called conservation biology, a discipline with its own journal and sense of itself. These scientists think of themselves as crash-cart biologists, the people who perform triage on the troubled life systems of the world. Their reason for being is a realization that human-caused disturbance of natural systems has already fostered a wave of extinctions worldwide on the order of the great extinctions that ended the Cretaceous period and ended the dinosaurs.

One of these scientists, Jared Diamond, speaks of the "evil quartet" of phenomena that is driving these extinctions. Three of them we hear about often—habitat loss and fragmentation and climate change—but the fourth does not slide so easily into our definition of evil, so is harder for the public and even scientists to swallow. The final piece of the evil quartet is invasion by exotic species.

In an editorial in *Conservation Biology*, Stanley Temple writes:

> Looking back through the pages of *Conservation Biology*, I am amazed to find so few articles on exotic organisms. . . . Alien species that have escaped the limits of their natural geographic

> ranges, as the result of intentional or inadvertent human activities, invade natural communities, with predictable consequences. Their predation, browsing and grazing, competition for limited resources, introduction of diseases and parasites, hybridization, and initiation of environmental chain reactions reduce natural diversity by causing extinctions and shifts in patterns of relative abundance. They should be a high-priority subject for research and action by conservation biologists, but apparently they aren't. Why? Conservation biologists should be as proficient at eradicating exotic species as they are at saving endangered species.

"Eradicating species": This is the notion that goes down so hard and cuts against the grain, the notion that spins the ambiguity into this topic. In his editorial, Temple shares a personal experience of killing trees to restore a native prairie and being criticized by environmentalists who couldn't imagine killing trees. Animal rights activists have stopped biologists from poisoning exotic rabbits that were extirpating a total of seventeen other species on one island. Animal rights activists protested the killing of exotic reindeer that had grazed islands in Alaska to near death.

We still have the gardening myth deep in our heads: that it is our God-given mission to line it up in rows and make it grow. We still have not got it into our heads that all of life is a complex of checks and balances, driven by both growing and dying, and that anything we do to tilt the system will spin through a place in waves of catastrophe for whole generations.

It is the greatest of ironies that Earth Day, a time that is supposed to enhance environmental awareness, is sometimes celebrated as an adjunct of Arbor Day. School kids, solemn-faced and righteous, plant trees, some of the very species of trees that Frank Meyer dug up in China to make the desert bloom. Arbor Day was invented and promoted by settlers in Nebraska who found the place "barren," probably Ralph Furnas and his friend J. Sterling Morton.

The school kids and disciples of Morton plant the same trees that the conservation biologists so assiduously eradicate today. The biologists have nothing against life but have come to understand that a life is at best meaningless and in worst cases massively destructive if it is lived out of its context. A weed is a plant out of place. We in the West have come to speak often of a sense of place, especially as we think about our human communities. This seems a good direction, but so far incomplete. It seems to me a more thorough understanding of this could come from knowing that plants are the sense of a place. The weather, evolution, the soils, and the fauna all mesh together in a shape we call community, and the manifestation of that community is plants as they grow together.

In a sense, this is no less true of the West now than it was before settlement. The plant life is the history of the place. Before settlement, it was a natural history, a record of wind, ice, and fire. We could find prickly pear cactus in the Black Hills and know the glaciers had come. We could find the seas of bunchgrass and tall grass and recognize the land of high-crowned teeth, meat eaters, and nomads.

But the tree people of Europe outran their native range and were pushed into the alien grass, the great desert. First they eradicated the native botany, swept the slate, then replanted with a botany of their devising. Now this place mostly grows exotic plants, the deliberate introductions in squared fields and straight rows, like convicts in forced labor in a pen. The accidental introductions fled the pens, crouching and scampering up ditch beds and roadbeds.

Because the roads follow the Jeffersonian grid, so do the weeds, and one can follow the path of settlement straight up the Missouri, out across the plains, and up the foot trails of the Rockies until the trails end, reading the place all the way, as it is written in Kentucky bluegrass and spotted knapweed. The grid was an abstraction, but the West has been remade with a plant face that wears it. This is how the history of settlement was written, and it may not be completely erased, even with the most virulent of formulations Dow Chemical can offer.

There is a specialty of science called palynology, the study of pollen. It turns out to be one of our most lucid windows into deep history. By taking core samples of ancient bogs or places where mud and pollen settled in layers, paleoecologists can read the circumstances of life a long way back through time. The type and relative amounts of pollen in a given layer are the signature of that time, telling whether a spot was forested, and with what sort of trees, whether it was grassed, and with what sort of grasses. From this, we may know the weather and fire history and interpolate the fauna.

Yet palynology is now a science only for historians, the paleo people. In the upper Midwest, where sites hold some of the richer pollen records, layers deposited since settlement are meaningless. The recent pollen records are a babble of species struggling to deny the conditions of the place. Palynological studies in 1993 examined the record at four widespread sites in the United States, stretching back for 5,500 years. At each site, the record was unmistakable; there were gradual waves of change until 150 years ago, and then upheaval came.

"Just in the last 100 to 150 years, grazing, logging, invasion by exotic species and the suppression of fires on which many ecosystems depend have brought about rates of ecological change unprecedented in their severity for the last 5,000 years," Dr. Kenneth L. Cole, the paleoecologist who directed the study, told *The New York Times*.

Allergists used to send patients particularly allergic to pollen to the arid Southwest to escape. The immigrants there used irrigation to plant bluegrass lawns and golf courses and exotic flowerbeds and orchards. Allergists now find it less useful to send their patients to the Southwest.

In 1993, a small army of botanists released the first installment of a project long needed but heretofore always too daunting to attempt. Essentially, the scientists are for the first time cataloging all of the plant life of North America, an estimated twenty thousand species of plants. The first two, issued in late 1993, of an expected fourteen volumes, the collective work of hundreds of botanists at thirty institutions, produced something of a shock for the botanists, which undoubtedly will be borne out as the subsequent twelve volumes unfold.

The catalog includes both native and exotic species, as it must in order to be complete. As a result of the work, scientists now realize that one fifth to one third of the number of species—as many as 6,600 species—in North America are exotic. The severity of this problem, the extinction of native plants it fosters, is not spread equally through the continent, however. The study found that the grasslands, the ecosystem that the study says covers 21–25 percent of Canada, the United States, and Mexico, suffered the most. In tall-grass prairie states like Iowa, Illinois, and Indiana, only 20 percent of the land remains sufficiently unaltered to be *potentially* capable of ever supporting natural plant communities. Grassland covers more land area than any other ecosystem in North America; no other system has suffered such a massive loss of life.

I am at poison school, the best name I can think to call it, spending a day with one hundred or so people, mostly men in ball caps in a meeting room outside of Missoula. This is a personal matter for each of us, this much we have in common. Each, including me, is licensed by the state to buy and apply on our own property the most toxic chemicals agribusiness has devised.

Most of the others in the room wear ball caps because they are ranchers. There's talk of livestock prices, the winter's weather so far. One of the few women in the room expresses a certain relief on seeing the lunch menu includes beef. She said she has only recently begun to recover from the outrage of attending a Cattlemen's Association meeting where chicken was served, an act not unlike serving a nice Chardonnay at a gathering of the Women's Christian Temperance Union.

A few in the room, however, are not ranchers, but people like me. We come to this convocation of chemical sprayers—nozzle heads, a friend calls them—the hard way. We come as environmentalists. In my personal care is 38 acres of mostly grassland, suffering greatly from knapweed, bluegrass, and especially leafy spurge. It was not long after discovering this that I found I was not alone. At cocktail parties in my town, even at those where Chardonnay is served, conversations about spurge have a way of drawing a crowd the way a discussion of a particularly lurid community scandal will draw a crowd at cocktail parties elsewhere.

Slowly, in these conversations, I underwent the same transition many other members of my community have undergone. *Silent Spring* was gospel to a generation of environmentalists, and we came to hate the chemical plague. Then some of us came to hate the plague of exotics even more, and we learned to spray Tordon.

As it is in most states, poison school in Montana is conducted by the local representatives of the land-grant college. The restoration ecologists in the crowd like me squirm and seethe during several of the day's lectures, which address topics such as why consumers should not be alarmed by pesticides and how the media has unduly inflamed them in this regard. It is not altogether clear what this had to do with our ability to handle pesticides safely, but the representative of the federal government still passes out the colorful and chirpy Dow Chemical Company brochures just to make sure the point took.

Halfway through the session, I start a conversation with a rancher, a man with a black Stetson and a pencil mustache forced to bridge some weathered crevices in his face. He has brought with him a volume of several hundred pages called *Western Weeds*, which he holds in his lap as a church lady holds a hymnal during prayer meeting. I have been meaning to buy this field guide, so I borrow it.

I open it, and he asks me to please not bend the binding like that. He tells me it is a thorough volume in that it lists all the weeds, both exotic and native. Like blue lupine. Lupine is a native legume, so a nitrogen fixer. It takes nitrogen from the air and puts it in the soil, so each lupine plant is a mini-fertilizer factory, only better. It doesn't leach nitrates.

I tell the cattleman I have some problem with calling any native plant a weed, but couldn't really blame him. Even as late as 1960, official USDA field guides listed "forb" as a synonym for "weed." He thinks about my assertion for a moment, then reminds me that cows can't eat some of the native plants, QED, they are weeds.

So I steer the conversation toward knapweed and spurge, subjects on which we can agree.

9

Roadside Attractions

It was one of those experiences meant to reinforce the illusion that all of the West is one big small town. The hour was a high one in the golden part of a July evening, and I had spent the twelve previous hours driving straight east across Montana; it takes that long. One simply puts the accelerator pedal to the floor and leaves it there. The trip crosses a whole time zone, yet we manage to fit the state comfortably into a single telephone area code.

I had deliberately chosen a route to emphasize my state's sparse population. One can cross Montana on the standard government-issue interstate that arcs like a maniacal grin across the southern third of the map. Many people do, especially in July. A century after settlement, the section-line roads that netted and tamed the West coagulated as super roads, and much of the baggage of motorized culture clotted along the new system of interstates. In summer, these arteries teem with shiny aluminum bodies like the Northwest's rivers that once ran with salmon before the dams came in. Winnebagos and mini-vans from the Midwest, families seeing a tourist's essence of the place. They follow block-typed chamber of commerce maps of the Olde West Trail. The Black Hills today, a day's drive, Yellowstone, a day's drive, Glacier, with stops at pullouts along the way for souvenir T-shirts and bumper stickers.

This coagulation has left some perfectly good highways, long, empty two-lanes that parallel the interstate, mostly vacant. State highway 200 and U.S. 2 across Montana—these I drove mostly alone, except for a few stops in cow towns, all the way into North Dakota.

Six hundred miles from home, I pulled into a parking lot at the Theodore Roosevelt National Park, and a uniformed ranger, complete with campaign hat and badge, walked up, scanned my license plates, and said, "You're from Missoula." Then he looked at me and called me by name. It turns out the ranger was a graduate student in Missoula, working on a master's degree in environmental studies. This was his summer job, and he knew me through some mutual friends and through some writing I had done.

Right off, he wanted me to write about the grasslands hereabouts, particularly the wildfire rush of hydrocarbon development that was turning a relict ecosystem into an oilfield. His plea was nothing new to anyone who writes about the West. People expect us to listen to their stories and help them beat back one more monster—mining, logging, dams, subdivisions. By now this exercise feel more like tag-team wrestling than journalism, and I am tired of the game, even as I realize its necessity. The monsters are real.

The West is made of one long series of necessary and true fill-in-the-blank stories, and sometimes it seems we are doomed to live them cyclically and perpetually, simply because there is no such thing as The Story. As the colonial culture of the West, we have no culture, which is just the same problem as having no story that tells us how we fit in the place. This is not an original idea, and in fact there is a self-conscious and active movement among western writers to invent a literature for the place. We need stories that will settle us to the land, not more stories reacting to those who would and do destroy it, but as long as the destruction goes on, these accounts of our struggles will be our only story. They are necessary, but seem doomed, a new sort of colonialism.

It is doomed because the West is a grassland. Literacy, literature, and self-conscious self-statement are creatures of civilization. Writing grew from commercial transactions and only subsequently evolved to religion and to literature, which through time has been wholly a creature of agrarian, settled societies. Mongol, Hun, and Blackfoot, the great nomadic grassland cultures, were always illiterate or anti-literate. Their stories were as mobile, ephemeral, and true as the rest of the landscape. Our story here has been one of ambivalence, which is to say, a test of faith.

During the settling of the plains, the reasonable god derived from the contracts of Christianity was beaten by an angrier and more jealous god. The plains' god was not the god of logic derived from thirty centuries of civilization but the god of fire and plague, a brutal and capricious creator like the predecessor god of the Christians, the Old Testament deity that had not yet consented to grace. The plains god was god as imagined by the Bedouin nomads of the grassland.

In his book *The First Dissident: The Book of Job in Today's Politics*, William Safire casts the story of Job in an interesting light, creating almost exactly the mirror image of the process that took place on the plains. Backing his argument with translations of pre-Christian versions of the biblical story, Safire declares that the story of Job and its subsequent reshaping in Judeo-Christian literature is a watershed, nothing less than a reshaping of the religion itself. At its root lies a fundamental contradiction between the human need for justice and the manifest injustice of the creation. The story's various permutations through the years only show religion's inability to resolve that contradiction despite its overwhelming need to do so. Our inability to derive a story for the grassland West is a subset of this failure.

For our purposes, it helps to understand the Joban tale's historical metamorphosis to its first written form. This is the best evidence that the way of life and worldview of the Jews and subsequently the Christians already had changed. The first written versions of Job are thought to date as far back as 2,900 years ago, about 800 B.C., with the best guess dating the story at about 500 B.C., but the folktale on which it was based was probably a thousand years older. Like the account of Cain and Abel, it arose in the patriarchal period when the Hebrews' way of life was that of nomadic herders.

The Book of Job is rife with contradiction and unresolved ambiguity, qualities that make it stand as literature, but are also the footprints of a story that traveled from its inventors, a nomadic grassland people, to the hands of civilized agrarians. In both versions—the oral traditions of the earlier nomads and the subsequent written versions of the civilized writers—Job is a good man who works hard and obeys the laws; he keeps his end of the contract, the covenant. Nonetheless, in a seemingly senseless act, God takes it all away and punishes Job. In earlier versions, Job is an outspoken dissident and critic of God. His arguments reveal a God of chaos, a God with whom we cannot reason. It is only in later versions that God fully reveals his reasons for this, his own battle with Satan, and, some scholars say, only in later versions is Job restored his former wealth. The priesthood, in need of a just God to enforce the rule of civilization, tacked on a happy ending. The story of the story since it was first written has been one of constant bowdlerization to emend and soften Job's blasphemy.

The recasting of the myth was necessary for a culture that was learning to follow the plow. In earlier days, the nomads could tolerate a capricious God; in fact, the shifting fortunes of the land demanded that very conclusion about the creation. God tells Job of the Behemoth, the Leviathan, a monster that is nature. It is God's job to keep the Leviathan in check, and to do so, he may need to work in mysterious ways. The central lesson of the precivilized versions of the story could be found in God's firm challenge to Job: "Where were you when I laid the earth's foundations?"

Listen up, Job, God is saying. There is something in the turning of the earth that your logic does not and cannot comprehend. Back off.

Later versions of the culture needed a God of reason, a God who would honor his covenant with Job, because later versions of the culture were agricultural. Farming is an activity that requires a fidelity to cycles, a contract that work will be done in spring in return for God's fulfillment of his contract each fall. It was the contract that Christianity had perfected by the late nineteenth century to a doctrine tuned to the needs of the hyper-civilization: industrialism. The contract had become the work ethic. Material blessings had become tangible evidence of God's love for the person so blessed. The settlers came to the plain flush with this notion, especially as it had been reinforced by nearly a century of Jeffersonian democracy. Make the desert bloom. Civilize wilderness. *Arbeit macht frei.* Imagine their surprise at meeting in this pagan land a pagan god, the Old Testament God, still alive and kicking.

The American grassland had for centuries been like all of the world's grasslands: nomadic, uncivilized, and therefore hostile to literature. Tents and constant motion make no allowances for libraries. Transactions were happenstance and did not produce the written contracts and receipts that elsewhere gave rise to writing. Still, it would be wrong to say that the American grassland did not produce a literature or that the literature it eventually did produce is false simply because it was an outgrowth of European settlement rather than an organic creation. The plains are as hostile to agriculture as they are to literature, but they nonetheless produced an agriculture. The hybrid literature the plains did produce is as relevant to our story as are field cultivators, center pivot irrigation systems, and barbed-wire fences.

Still, all of these things have the air of imposition surrounding them. In the literature, this shows as an ambivalence. The literature is a record of the crash of European culture against the hard face of the place, just as the story of Job was the crash of the human need for justice against creation's overarching injustice.

This theme plays strongest in *Giants in the Earth*, O. E. Rölvaag's 1927 novel about Scandinavian settlement in the northern plains. We get the best view of what the story will become in the title of his climactic chapter: "The Great Plain Drinks the Blood of Christian Men and Is Satisfied." Settlement was a test of faith. Later novels turn this ambivalence to subtler symbols. The religious overtones fade but do not disappear in the works of such writers as Mildred Walker, Mari Sandoz, Wallace Stegner, Willa Cather, and recent works such as Jane Smiley's *A Thousand Acres*. Rölvaag sets up the framework for this thread of ambivalence in two character types: the practical plodder who in exchange for full citizenship on the plains surrenders his humanity to the demands of the place and the wistful romantic who will forever be a misfit in the place.

In a real sense, these are the two halves of Job's character, but this ambivalence is the personal story of each of the writers. Within each writer are both of the seminal character types of their work. Rölvaag, Sandoz, Stegner, and Cather, for instance, all wrote fondly of the plains but lived most of their working lives elsewhere. Stegner lived in California, Cather and Sandoz in New York. Rölvaag returned periodically to his native Norway. Far from undermining their stories' credibility, however, their leaving their homes is what makes their stories real. It was and is a hard place for literature to inhabit.

In *Giants in the Earth*, the ambivalence is borne in the relationship between Per Hansa and his wife, Beret, the couple that immigrates from Norway to settle in South Dakota. Beret is the book's dark presence, the character who suffers the plains, but just as she seems to shrink, wither, and sadden under the strain of plains life, Per seems swelled by the challenges of their Red River homestead, but he needs to grow to match the epic battle. He needs to become a giant, as Rölvaag reveals in a quote from the Book of Genesis that produces the title of the novel:

> There were giants in the earth in those days; and also after that, when the sons of God came into the daughters of men, and they bore children to them, the same became mighty men which were of old, men of renown.

Or at least that's how they started. Those were the yeomen's marching orders, a charge Rölvaag later echoes in even more explicit language, which echoes the Joban charge:

> And now he [Per] had begun a seemingly endless struggle between man's fortitude in adversity, on the one hand, and evil in high places, on the other.

Settlement was not simply the plowing of dirt, but battling the forces of evil, which was nature as Leviathan. Beret, however, sees the battle lines less clearly, and periodically the novel shows us Per's metamorphosis from another angle. To Beret, Per was not so much becoming a giant pitched against evil as he was becoming a man lessened by infection with the manifest evil of the place. Of those people adapting to the hardship of the plains, Rölvaag says: "Everything human in them would be gradually blotted out." Beret says of Per:

> Couldn't he understand that if the Lord God had intended these infinities to be peopled, He would not have left them desolate down through all the ages . . . until now, when the end was nearing?

One may hear in Rölvaag's words and in those he gives Beret the epithet "barbarian," this time applied to civilized people adapting to grassland ways. Beret sees this transformation as Per's loss of faith. A new God had replaced the gentle and protecting God of her safe village in Norway. Conveniently, there existed on the plains an Old Testament–style plague that writers like Rölvaag could use when they wished to invoke this angry God. The hordes of grasshoppers (read: plague of locusts) were real enough events for those early settlers, and in some places still are. It is not at all far-fetched to suggest they are and were retribution from an angry creation. There is some suggestion that the overgrazing of the mid-grass prairie, already well under way when the yeomen hit the region, had allowed sunlight to penetrate to the surface of the soil to incubate grasshopper eggs. This, in turn, allowed unheard of numbers of grasshoppers to hatch and form the clouds of locusts.

Or maybe not. Maybe these outbreaks are just something that happens every now and again, but Rölvaag wasn't equipped to see it that way. He wrote:

> And now from out the sky gushed down with cruel force a living, pulsating stream, striking the backs of the helpless folks like pebbles thrown by an unseen hand; but that which fell out of the heavens was not pebbles, nor rain drops, nor hail, for then it would have lain inanimate where it fell; this substance had no sooner fallen than it popped up again, crackling and snapping —rose up and disappeared in the twinkling of an eye; it flared and flittered around them like light gone mad; it chirped and buzzed through the air; it snapped and hopped along the ground; the whole place was a weltering turmoil of raging little demons; if one looked for a moment into the wind, one saw nothing but glittering lightning-like flashes—flashes that came and went, in the heart of a cloud made up of innumerable dark-brown clicking bodies! All the while the roaring sound continued. . . . All that grew above the ground, with the exception of the wild grass, it would pounce upon and destroy; the grass it left untouched because it had grown here ere time was and *without the aid of man's hand* [italics added].

The grass was the creation and favored child of the weird god of this weird place and only it could survive. The grasshoppers were the heavens gone mad. As the novel trudges through blizzard, crop failure, plagues, wind, and more blizzard, it becomes clear that in Christian eyes the land itself is madness that offers bad ends for Christian men. For Per, who adapted and willingly surrendered his faith and civilized ways, his sin is hubris, for which he pays by freezing to death in a blizzard. For Beret, there is sadness and madness. Says Rölvaag:

> But more to be dreaded than this tribulation was the strange spell of sadness which the unbroken solitude cast upon the minds of some. Many took their own lives; asylum after asylum was filled with disordered beings who had once been human. It is hard for the eye to wander from sky line to sky line, year in and year out, without finding a resting place!

Mildred Walker's 1944 novel *Winter Wheat* sets up almost exactly the same dichotomy between husband and wife, a wistful dreamer and a practical adapter running a wheat farm in Montana. In this case, however, the roles are reversed and it is the wife (seen through the eyes of her daughter and the narrator) who adapts and loses the faith taught to her in her native Russia. Again the relentless space swallows the European God. After a city boy suitor, Gil, jilts the daughter, partly because of her coarse prairie ways, she says:

> I wondered once if it would have done any good if I had prayed to keep Gil. Then I looked up at the endless blue sky reaching way beyond the pale outlines of the mountain, over the stubble and the wheat. It seemed too big to pray to. We three looked too small and unimportant down here on the ground. And there were the Yonkos and the Bradiches and the Hakkulas and the Halvorsens, all out threshing too. If we were all saying prayers, one of us asking for one thing and someone else asking for the very opposite, we would sound like the hungry squealing of the pigs.

Throughout prairie literature, the landscape is the rock on which European pieties founder, a theme especially prominent in

Willa Cather's work. Many of her stories feature something of a stock character, a cultured European somehow stuck on the plains and there lost. Like Spanish Johnny, Cather usually gives these people command of music or sometimes books, and like Johnny they meet hard deaths, more often than not suicides, leaving behind a treasured fiddle or a fine-bound book. Cather's prairie kills romance, music, and poetry, but her stories also resurrect these qualities in new characters: children of the prairie who learn the quiet and sad poetry of the land so well that they must leave it. Like Cather, these children head for the city to write what they know.

This scheme plays out most particularly in her best-known novel, *My Ántonia*. The novel first gives us Europe in Mr. Shimerda, a Czech emigrant, a violinist, and eventually a suicide. His death leaves the story to be borne by two children, his daughter, Ántonia, and Jim, a neighbor boy, our narrator and Cather's alter ego.

Jim and Ántonia are archtypical children of prairie literature, similar to those in Walker's work, in Sandoz's, and even in Rölvaag's. Their reality rings even in *Wolf Willow*, Stegner's story of his childhood in southern Saskatchewan. Stegner writes that the place made of him a "sensuous little savage" and provided a "childhood of freedom":

> We had our own grain and our knots as well, but prairie and town did the shaping, and sometimes I have wondered if they did not cut us to a pattern no longer viable. Far more than Henry Adams, I have felt myself entitled to ask whether my needs and my education were not ludicrously out of phase. Not because I was educated for the past instead of the future—most education trains us for the past, as most preparation for war readies us for the war just over—but because I was educated for the wrong place. Education tried, inadequately and hopelessly, to make a European out of me.

Freedom, too, is the word for the childhood of Jim and Ántonia, as if childhood was the new incarnation of the romantic notion of noble savage. As in "Spanish Johnny," the prairie is almost universally depicted as having "hands gentle to a child," as if the state of nature is appreciated only by the unschooled and unspoiled minds of children and Indians, the sensuous little savages.

Accordingly, the literature often is cast as coming-of-age stories: as a loss of innocence shattered by the harsh life not of the plains, but of the adult work of taming the plains. The "hands so gentle to a child had killed so many men." It is in this process of coming of age that the ambivalence key to other prairie literature reemerges. Education does make a European of sorts out of Jim, and he leaves for the cities of the East Coast. Ántonia, meanwhile, surrenders her innocence and romance, even her humanity, to become the practical adapter, a drudge of a farmwife married to a drudge of a farmer. Remarkably, though not at all remarkable if one understands the reality of grassland life, the metaphor that emerges in both Stegner's and Cather's work to explain this process of taming is the roads; the likes of Spanish Johnny could survive only "before the Road came in."

The first "roads," or at least the basis of them, existed before Europeans ever saw the plains. The grassland is ruled by motion, and all of its inhabitants learn to move or die. Through most of the grassland, this motion requires no roads. It is a place of freedom, and one could always and in some places still can simply pick a point on the horizon and walk there without benefit of pre-trod path.

There emerged, however—through certain draws, in hills, in brushy river valleys—paths threaded through by the inhabitants. Most of these meanders were blazed by the buffalo. They flowed with the landscape as it was read by bison, the same way a river reads the contours of its valleys.

At first, humans followed these roads. Especially east of the Mississippi, where the wooded landscape made trails more necessary, buffalo were the early guides of travel, first for Indians, then settlers. Buffalo trails were the basis of the Wilderness Trail from

Virginia through the Cumberland Gap to Kentucky. The rail line over the most feasible portage between the Potomac and Ohio rivers simply followed an old bison trail. A buffalo trail up the Platte River became the route of the Union Pacific Railroad. Centuries of travel had taught the animals the best way, and the railroad's engineers simply followed it.

The first settlers in the West rolled their wagons across the prairie, and the steel-rimmed wheels announced the coming of a new age. Rölvaag catches this bit of news as Per Hansa's two wagons enter the South Dakota prairie:

> Both wagons creaked and groaned loudly every time they bounced over a tussock or hove out of a hollow. . . . "Squeak, squeak!" said one. . . . "Squeak, squeak!" answered the other. . . . The strident sound broke the silence of centuries.

The wheels shrieked the news that the grassland's rules of free motion, like the freedom of children, was to be irrevocably changed. Stegner picks up the news in *Wolf Willow*:

> And that is why I so loved the trails and paths we made. They were ceremonial, an insistence not only that we had a right to be in sight on the prairie but that we owned and controlled a piece of it. In a country practically without landmarks, as that part of Saskatchewan was, it might have been assumed that any road would be a comfort to the soul. But I don't recall feeling anything special about the graded road that led us more than half of the way from town to homestead, or for the wiggling tracks that turned off to the homesteads of all others. It was our own trail lightly worn, its ruts a slightly fresher green where old cured grass had been rubbed away, that lifted my heart. It took off across the prairie like an extension of myself. Our own wheels had made it: broad, iron-shod wagon wheels first, then narrow democrat wheels that cut through the mat of grass and scored the earth until it blew and washed and started a rut, then

> finally the wheels of the Ford. . . . Here is the pioneer root-cause of the American cult of Progress, the satisfaction that *Homo fabricans* feels in altering to his own purposes the virgin earth. Those tracks demonstrated our existence as triumphantly as an Indian is demonstrated by his handprint in ochre on a cliff wall.

Stegner speaks of two sorts of roads and weighs in with a clear preference for the meander of wagon tracks on his own land. That's how the early settler roads began, just another phase of the paths that had woven the place together for millennia. But as settlement accreted, the rectilinear cadastral grid came into play. The roads were deliberately moved to the lines of the grid. They snapped to the section lines, because the section lines were boundaries between the homesteads, the logical place to locate rights of way. The grid became what is almost universally the pattern of grassland roads today: the section-line roads run straight and exactly a mile apart, north and south, east and west.

I think Cather means to tell us about the significance of this process in the real taming of the place. When Mr. Shimerda kills himself, the neighbors bury him out in the open prairie, near the boundary of his land. Only later, when the section-line roads come in, does this become a problem, and then the string-straight section-line road must jog a bit, just to miss the grave on the line. His grave becomes a sort of memorial to the cost of the taming.

At the end of the novel, the cycle of taming is reiterated for us when Jim, after not seeing Ántonia for most of their adult lives, finally meets her as an aging farmwife. This reconciliation serves as a late meeting of the two halves of prairie character, and Cather weaves it together with a road. Jim is getting to know Ántonia's children, still wild things, as he once was, who roll and frolic with the prairie. In the midst of this scene, he stumbles on the remnants of an old, abandoned meandering road. He says:

> Everywhere else it had been ploughed under when the highways were surveyed; this half mile or so within the pasture fence was

> all that was left of that old road which used to run like a wild thing across the open prairie, cling to the high places and circling and doubling like a rabbit before the hounds.

Tame roads, children tamed as adults—all are the same.

All the way from the Rocky Mountains, the old seabed that has been carved and rolled to the plains is, close up, anything but regular. But still it is a plain and on the whole regular enough to earn the name. In each mile to the east, it slopes ten feet, clear to the Mississippi. This slope is responsible for so much of what the grassland is. It means that the water will move east, the land will drain, but slowly. When the grass is alive and well, most of the rain will be filtered and caught to seep beneath the surface and there build the life of the place, its nature.

But the tilt still is a drain, as Sandoz says, the drain that allowed the nature, the riches, of the place to be bled off and sent east. Then the railroads came and then the roads came, all following the east-west lines the surveyors had cut, and the drain accelerates, now in some places no longer impeded by the grasping roots of the grass that have been put to the plow.

Jane Smiley's 1992 novel *A Thousand Acres* is regarded, and rightly in a sense, as a break in our literature's treatment of American farmers. The Iowa farm family is anything but the noble yeomen Jefferson and Crèvecoeur envisioned. Where the yeoman myth postulated freedom and independence, Smiley gives us alcoholism, incest, and alienation. The social statistics of rural America go a long way toward endorsing Smiley's view. It is tempting here to say that the yeoman myth so seized our literature that it took two hundred years for literature to fly straight in its face, as Smiley has done.

But in a more important sense, Smiley is still in concert with her predecessors. The brutal and dehumanized patriarch of Smiley's Cook family is only a modern incarnation of Per Hansa or especially of Sandoz's own father, Jules Sandoz, who was the topic of *Old Jules*, her first book and one of the most remarkable works of plains literature. Smiley again parallels the stripping of nature from the land and the stripping of humanity from its inhabitants.

Her metaphor for this is again transportation; she first calls them "lines," then we learn they are tile lines, the drainage tiles that underlay the Iowa fields and allowed them to become productive. The story of those tiles on flatlands near the Mississippi is as much a story of settlement as the roads were farther West in the drier reaches of the prairie, but because the tile fields are subterranean, they are unseen, a condition broken only occasionally by the catchment basins that breach the surface.

Beneath runs the water, the symbol that for Smiley comes to stand for the nature of the place. It is this water, drained, that created the wealth of the farmers. Their fields once were wetlands and couldn't be farmed before the tiles were dug in. This same water, however, causes the miscarriages of Ginny, the story's narrator, because of the nitrate pollution it carries from chemical fertilizers, the new nature of the place.

To the farmers, especially the men who populate Smiley's novel, the water is invisible, as forgotten as the prairie marshlands it once supported. The fields are flat, square, and grow corn, but the water is visible to Ginny through memory, her memories of herself as a child. In this, she is the same child that appears in the work of the rest of the prairie authors. She has the same child's wisdom of nature. Ginny tells us:

> I was always aware, I think, of the water in the soil, the way it travels from particle to particle, molecules adhering, clustering, evaporating, heating, cooling, freezing, rising upward to the surface and fogging the cool air or sinking downward, dissolving this nutrient and that, quick in everything it does, endlessly working and flowing, a river sometimes, a lake sometimes. When I was very young, I imagined it ready at any time to rise and cover the earth again, except for the tile lines. Prairie settlers always saw a sea or an ocean of grass, could never think of any other metaphor, since most of them had lately seen the Atlantic.

The Davises did find a shimmering sheet punctuated by cattails and sweet flax. The grass is gone, now, and the marshes, "the big wet prairie," but the sea is still beneath our feet and we walk on it.

Settlement was a subterranean battle for the life of the place. Here we have now a child's promise that it may rise again, but a child's mind cannot comprehend just how much damage was done.

We who would inhabit the grassland need a new story, a sort of illiterature that rises from the land. I have no idea what that might look like as a whole, only some clues. I keep coming back to the Hornaday expedition that collected the last few bison from the upper Yellowstone for taxidermy. On my way to Teddy Roosevelt's park, I had spent the last few hours traveling through the very strip of land where this hunt had occurred in 1886. It's still a harsh place of short grass and hard-baked hills worked to brittleness by cattle and sheep. The towns like Jordan and Circle are spaced an hour apart with nothing but vacant road between, a pleasingly lonely quarter of the universe, and in town, people still savor the luxury of this loneliness. The towns do not dress much beyond dirt sidewalks, faded storefronts, and Coca-Cola signs decades out of date. The markets pack pickup truck loads of groceries in boxes and tie them up with string. The ranchers pay with counter checks. Handbills announce auctions for pedigreed bulls and rodeos. There are no bison.

One hundred seven years earlier, Hornaday misread a story here, a statement as scathing, eloquent, and angry as any written here since. Remember how he described in his journal the condition of his party's buffalo kill after the Indians had taken the skin and meat during the night: "Through laziness they had left the head unskinned, but on the one side of it they had smeared the hair with red war-paint, and the other side they had daubed with yellow and around the base of one horn they had tied a strip of red flannel as a signal of defiance."

If there ever arises a true story of this place, it will be written in war paint and strips of red flannel. It will be stored in the stout library containing generations of prairie information, stored as the living soil.

While driving through the landscape, I had been listening to a tape of the poet Gary Snyder reading about stories. He said the land calls forth stories. Some landscapes call forth trees and some call for grass or for whales. If this is so, then we need to wonder why the Badlands has called forth the story that it wears.

Badlands is a term that crops up throughout the West and is generally applied to any heavily eroded system in the midst of otherwise gentle plain. The most famous example, and the one that gets to wear the term as a proper name, is the strip along the Little Missouri River that parallels the Montana–North Dakota boundary just on the east side of the line. At its center lies the Theodore Roosevelt National Park, which plays out in two dollops, one along the interstate highway around the town of Medora and one maybe fifty miles to the north. These poles are surrounded by the Little Missouri National Grasslands and together make up most of the public lands of North Dakota.

On first view, this place does indeed look bad as hell. Early cowboys called it "hell with the fires put out." They weren't wrong. Like most buttes, the Badlands were formed by the deposition of harder materials over soft layers and subsequent undercutting and erosion. The hard rock for the Badlands moved in, carried on water, by streams from the forming Rocky Mountains a hard day's drive west. Most of this happened about the same time the Rockies were raising high-crowned teeth on horses, more than twenty-five million years ago. In this particular butte landscape, though, there was a twist in the scheme. Before the rising mountains had called forth this place's aridity, this land had sponsored a lush coat of vegetation, now dead stuff layered in the earth as coal. Lightning fired the coal to create an in situ kiln that roasted some of the sediment layers of clay red and brittle like Indian pots. Now, with the fires put out, the place still can't seem to shake the look of a place where hell has been.

More than looks though, it is, to its rocky bone, a brash, tough place, so humans have mostly left it alone. It was not even the white farmers who were unable to settle this place who made up its name. The Sioux called it *Mako shika,* the "no good land."

This string of vituperative labels has been the land's saving grace, allowing it to stand as a sort of wild Gibraltar in a flat sea of agriculture. Just to the east the sea breaks in earnest, and immigrants raise wheat. In a wheat field down the road from the Saint Demetrius Ukrainian Catholic Church, a white wood building no different from all the other churches of the high plains except for its onion dome, the farmers have raised a monument to themselves on a bit of a sodded hill. The sign under the white cross says the settlers came at the turn of a century and "within the span of a lifetime, they developed here in Dakota a farming empire undreamed of by man."

Head east from the Badlands a day across the flatland, and the scene changes only in the details. The white churches segue from Ukrainian to Scandinavian or German Lutheran, the wheat to soybeans and corn—all the way to Minneapolis and Fargo.

As this monoculture of industrial agriculture was unrolling across the plains in the nineteenth century, probably no one had a clue as to where it was headed, but if anyone did, it was the young Theodore Roosevelt. To some, he was the father of the modern conservation movement, and there can be no doubt his contributions in this regard were enormous. Yet more exactly, he was the first figurehead of a wing of the movement, the blueblood, patrician conservationist born of ambivalence. Industrialism of the late nineteenth century was progress to the bluebloods of the time, the Victorians, and they were all for progress, but they knew it came at a cost. Part of blunting industrialism's blow meant setting aside national parks and wildlife preserves. The same instinct led to child labor laws, a recognition that our way of work and of doing business had become sufficiently brutal to make it necessary to shield wild animals and small children from it.

Yet to see the Badlands is to imagine there is more to Roosevelt's conservation than a vision of natural playgrounds for the rich. Behind the national park's visitor center at Medora, there stands a hand-hewn log cabin moved here from its original site, a ranch on the Little Missouri just to the south. It is the Maltese Cross cabin and was Roosevelt's first home here when he visited the Badlands in September of 1883. Some of the original bison remained and he hunted them, but mostly he roamed the place, a young man from New York expanding his spirit to match the scope of the surroundings. The following winter he returned to the East, and there watched both his mother and wife die within hours of each other. The following summer he returned to the Maltese Cross cabin, and it became a place of grief.

It is just a building now, meticulously restored and conveniently perched at the roadside in range of video cameras, a safe walk even in white shoes from the visitor's center blacktop parking lot. It's just a building moved from its place and so, looking at it, one can't tell what that place must have been to Roosevelt, a dry, rocky stretch of badlands, buttes, baked red rocks, bison, birds, and snakes, a place of grief that built something in him we can only try to read now in the sediments and deposits of a century's time.

"I never would have been President if it had not been for my experiences in North Dakota," he wrote later.

When Roosevelt was president, the fight to set aside public lands headed his agenda. His Progressive sidekick was Gifford Pinchot, a fellow patrician and the first head of the U.S. Forest Service. They consolidated the system of forest reserves begun in 1891 into the Forest Service. When a hostile Congress passed legislation forbidding expansion of the system, Roosevelt sat up all night with maps and did exactly that before the law could take effect.

Then, though, this reserve was a forest system in that conservation was practiced in forested land and lands never settled. It would take another generation and the disaster of the dust bowl not just to stay but reverse some of the worst effects of settlement. During the height of the Depression, when marginal and even less-than-marginal agricultural lands were becoming airborne, Congress passed the Bankhead-Jones Farm Tenant Act, which empowered the federal government to buy back cropped land from busted farmers and preserve and revegetate it with grasses as public reserves for soil conservation. Initially management of these lands fell to the new Soil Conservation Service, later to the Forest Service.

Today there are seventeen national grasslands, encompassing a total of just more than 4 million acres within their boundaries. Of these lands, 3.75 million acres are publicly owned; the rest are private inholdings: land of farmers and ranchers who didn't sell but whose ranches lie within the boundaries of the public lands. Most of the national grasslands are spread through the West, one in the Pacific Northwest. More than a fourth of the total acreage lies altogether in the Little Missouri National Grasslands, which is to say, the North Dakota Badlands. The land set aside as the Theodore Roosevelt National Park is two islands, a north and a south unit managed by the National Park Service. Surrounding the islands, stretching from Williston on the north to Marmarth on the south, runs the 120-mile-long band of public lands, a grassland run by the Forest Service.

A look at a map of the place color-coded to indicate ownership shows right away that it is no band. Congress meant for the Bankhead-Jones lands to serve as sorts of buffers to the soil erosion and abuses of adjacent farmlands and as demonstration projects. To meet both goals, Congress decided the new public lands ought to be interspersed with private ranches. The Little Missouri National Grasslands reflect that goal as well as the older legacy of the rectilinear cadastral survey. Squared private holdings interlock with squared public holdings to form a checkerboard, as it is always called, but a chessboard would be a better metaphor. Some of the players have more power than others.

Seeing only the gross boundaries of this place but not the shading that reveals its checkered past tempts one to imagine it as an opportunity for the restoration of a large island of grassland ecosystem. The two pieces of Roosevelt's namesake park are epicenters of resurrection radiating to the rolling plains and gulches of the surrounding national grasslands. This notion is given at least lip service by the Forest Service, Pinchot's heirs, who lately espouse as their latest new idea something called "ecosystem management." Generally, this idea suggests the keepers of public land ought to look at the whole of the place, then come up with a plan of "management" matched to the climactic, geologic, and biological contours. A piece of land, though, has history and sometimes ecological ideals bang hard against that legacy.

Speaking with Sam Redfern, the Forest Service's district ranger and nominally the man in charge of enforcing this ideal, one believes that, unlike some other Forest Service managers, he actually would like to do right by the place, not just turn a buck on timber sales and oil, mineral, and grazing leases. He's maybe forty, thin, and bald, and speaks directly and informally.

The problem is, he is only nominally in charge.

The two islands of national park in the grasslands raise bison, bighorn sheep, elk, mule deer, and antelope. The first three all were extinct in the region and had to be restored through some human meddling. The sheep are California bighorns introduced in 1956; the plains species, the Audubon bighorn, is extinct.

It is the bison, though, that catch my eye, mostly because they are the manifestation of a healthy native grassland the way a flower is a manifestation of a healthy plant. The night before I talked to Sam, I had pitched my tent in a grassy draw in the park, and one large bull, pumped up by the approaching season of rut, had grumbled and roared through sunrise all around my camp, pawing, rolling in the dust, and hooking out a juniper with one of his horns. During the interview, bison were still on my mind and a place called a "national grassland" seemed like it ought to be their home, especially one honoring a system called "ecosystem management."

No, said Sam, the bison are not allowed to wander outside the islands of park onto the surrounding Forest Service grassland. The park is fenced like a stalag. Grass is for cows. The ranchers who hold the grazing allotments on the public land and own the private land will not tolerate bison or for that matter elk straying from the park. By law this decision is the ranchers' to make. The Forest Service is manager of the grassland only formally. Under Bankhead-Jones, the authority to set grazing allotments, decide the terms of the leases, and, in effect, manage the grasslands falls to associations made up of the ranchers who hold the leases. The government's grassland is a cattle range, a band of public lands dedicated to the support of a quarter of a million cattle.

Anyway, it is not certain the range could support bison. Owing to the history of the place, some of the grasslands are not native grasses. Sam says that before the federal government bought it from dust bowl refugees, much of the range had been plowed and cropped by yeoman farmers. The Badlands are marginal lands, though, so these settlers were among the first to go belly up when the hard times hit and the first to be bought out. The federal government then didn't know much about the integrity of native ecosystems, so it reseeded the lands to crested wheat grass, the exotic species largely worthless to wildlife and unable to reestablish the web or relationships that build a healthy plant community.

"Once they turn the soil over, it's like mining it," says Sam. "It will never be the same."

To round out the matter, the exotic plague of a weed, leafy spurge, is working its way everywhere, wiping out entire draws and ridges, even in the park, but mostly up the roads. There are five hundred miles of new roads in the past few years (another story wound altogether in the political history and deep geological history that gave the place oil). As one drives the grassland, there is no illusion of traveling a pristine natural area or even an open range. Plain and simple, the Little Missouri National Grasslands is an oilfield and likely will remain one for the course of a young person's lifetime. Not much can be done about this; the drill rigs will leave only when the oil runs out. The federal government's purchase of this land in the Dirty Thirties was only a reclaiming of the place. It, of course, originally belonged to the government, but much of it became part of grants the government gave the railroads as an inducement to stretch their lines to meet Manifest Destiny. The railroads sold the land to the farmers but kept the mineral rights. Those rights followed separate paths of ownership and now are in private hands; the public acquired only surface rights when it bought out the bankrupt farmers.

The grasslands are part of the Williston Basin, an area recently found to be heavily underlain with oil. Steve Cohen, the park ranger who recognized me in the parking lot, tells me that sometimes on a summer's night one can climb a butte in the park and survey the surrounding landscape of prairie and see the oil wells flare like candles.

"It looks like Kuwait," he says.

Legally, says Sam, there's nothing the federal government could do about this. Under our hydrocarbon society's system of laws, a mineral right supersedes a surface owner's right to stop development. There are already about three hundred oil and gas wells on the Little Missouri National Grasslands. The Forest Service is completing a study of leasing those lands to which the public holds the mineral rights and likely will allow further development.

Sam says that unlike oil drilling on fragile mountain slopes, these grasslands heal easily, so one can reclaim a drill site and the road that led to it. Within the last few years, when the oil companies built those five hundred new miles of roads, the Forest Service closed and reclaimed one hundred miles with native species of plants. The prairie does heal and in the long run oil development likely will leave fewer tracks than farming has done, with its miles of crested wheat grass and spurge.

But it is a long run.

"They're still finding oil, and it still makes money producing it," Sam says.

• •

The Badlands are again making a young man from New York City plot out the course of his life. Steve Cohen landed here by accident more or less and the place has caught his imagination. He has worked as a seasonal ranger for two summers while finishing his degree. He's not sure yet whether he will become an environmental activist or a schoolteacher specializing in environmental science, and sometimes it is not clear there is a difference.

For now, though, he would just like to see people come to appreciate this place, which is why he works here as a ranger. He leads interpretive tours of hikers every day through the wilder stretches of the park, teaching people to name the place, to read the place. A few get it right away, and so he's come to believe his work has merit. He takes maybe a dozen people a day on his walking tour. Meanwhile, on a busy summer's day, one thousand people will drive through the park's gates, cruise the blacktop loops, pull out at the blacktop scenic overlooks, videotape the sign and the prairie dogs, and drive on.

Even from the center of the park's more remote reaches, visitors can hear the exhaust of the oil rigs on the surrounding grasslands. Sometimes they complain about this. Sometimes the complainers are driving Winnebagos that get six miles to the gallon of gas from this earth.

Steve is telling me all of this as we sit on the deck of a bar in Medora. Now and again, though, we have to stop our conversation. The Burlington Northern freights rumble and shake through town, obliterating anything we might say.

Up north of the park is a stretch of land that was Theodore Roosevelt's ranch, the Elkhorn, the spread he bought after his time in the Maltese Cross cabin. Actually, following the fashion of the day, Roosevelt really bought only the buildings. The land was public. Roosevelt went broke in the cattle business during one of the big die-ups around the turn of the century and wound up giving the ranch to some employees. It went to ruins and eventually became the object of an archaeological dig, just like the Mandan and Arikara ruins hereabouts. Today, if you drive toward it through the grasslands, not the park, it's easy to spot. There is a sign that says "Elkhorn Ranch Oilfield." I wonder what TR, the trust-buster, would make of this epitaph to his many political battles with big oil.

In the Maltese Cross cabin and in the surrounding Badlands, Roosevelt read something, a literature of the land that settled his grief and shaped his life. Properly considered, the story of this land created much of the conservation law our nation enjoys today. What Roosevelt learned from this place, the ineffable sense of the wild, is what he took to the presidency. But what of Steve Cohen's time? Can this story that Roosevelt saw be read from Winnebagos and recorded with video cameras? And what story is there to read? Once this land called forth bison and wind as its story. Now it calls forth only oil wells. This is our modern hell, with plenty of oil to fuel its fires. And there must be plenty of oil, because this is a hell of roads.

Just south of the Badlands one enters the Black Hills, a preserve of roadside attractions. Toward the center, there is a herd of bison mired in a fenced compound at the roadside for the viewing. Thousands of people stop to do so.

Just to the south of this lies Wind Cave National Park, one of the nation's best examples of a restored mixed-grass prairie. In it one can walk off in any direction, just pick a point on the horizon and go, as I have done, the only time in my long history of just plain walking that walking has felt like flight. In late summer, one walks among bronzed stalks of big bluestem and sees instantly why the first settlers called it "turkey foot." Little bluestem, side oats. Leadplant. Elk bugle in the evening. Topping a rise brings one face to face with a sulking bison bull. Sometimes an antelope will parallel a hiker's course, as if it is just going along for the walk, as antelope have done longer than any large mammal on this continent. Sit in prairie dog towns and watch for burrowing owls. Mind the snakes.

To do all this, one must only brave the visitor center, where a jammed parking lot suggests the walk may be crowded, but it's not. The people are all there to see the limestone caves that give the park its name. Almost no one ventures beyond the parking lot. The rangers became excited when I asked for a back-country permit. People don't often make that request, and the rangers had to rummage through a pile of papers for the little-used file of permits.

Between 1983 and the end of the decade, back-country use at the nation's parks was cut in half, while the number of front-country visitors, the drive-throughs, rose steadily.

Thus we close the circle. We found the American West a curious place, alien and bare to our eyes. Because of this, we failed to allow it to tell us its story, to give us its name. We failed to learn from the plants and the people who knew them as a way of life. We failed to regard the animals for what they were and wiped all of this out to replace it with a world of our own devising. Because we imported this world, it came in on the roads we built for it and flourished at the roadside the way brush does along a desert stream. And now, absent from nature and feeling a vague sense of the loss, we return to the West, increasingly along the same section-line roads, and believe that when we see a pen full of bison we are seeing nature. What we see at the roadside is not nature, but a face we have painted for nature. The leafy spurge, crested wheat grass, and penned bison are our own images reflected through a fence.

We videotape them and we drive on.

10

A Place's Assertion

The whole time I have been writing this book there have stood in a small clay vase on a shelf above my desk five stalks of prairie sandreed grass. The species can be found in Montana, but it is rare here. One is more likely to find them farther east, in mixed-grass prairie, in sand soils, particularly in the Sandhills of Nebraska, where I collected my five stalks, brown and desiccated on that winter day, as they are now. They came from a particular grave in a place full of history, but that alone is not why I keep them where they are. They are for me voices of a future, testimony of the power of plants to assert themselves against artifact and to resurrect the nature of a place.

The Sandhills are the largest set of sand dunes in the Western Hemisphere and cover most of Nebraska north of the Platte River. They are simply very young mountains of sand whipped to a froth by powerful winds. The sand slowly gathered a coat of specialized grasses. These sealed the surface long enough to begin grass's cycle of soil-building, which in turn created a base for the more fastidious species of the mixed-grass prairies. The grass mostly held, except for formations known as blowouts, hillside depressions where the wind manages to pry loose a chunk of sod and pick at the sore as a child picks at a scab. Blowouts still appear here and there around the place like wind-posted warning signs of the fragility of the post-glacial contract.

During settlement, however, these postings were ignored, and the round hills were cut and sold as 160-acre rectangles. We know much about this process in the Sandhills because many of those settlers had a difficult time finding their allotted 160 acres, especially because the cattlemen already in the vicinity had other ideas for the land. Open shooting wars developed, as they did throughout the cowmen's West. A central figure in this was a Swiss immigrant who was a homestead locator, Jules Sandoz.

His life may be summarized in a single and signal act. Once, when he heard of the cattlemen's threats against his life, he simply went to a nearby ranch, the Spade, and staged a public target practice so that the cowboys might come to better appreciate his considerable skills with a rifle. He is known for bringing settlers and orchards into the country. He is known especially because he is the subject of *Old Jules*, written and rewritten by his daughter Mari Sandoz, until *The Atlantic Monthly* printed a part of it in 1935 and made for her a career away from the plains. Often she returned, though, and she did so finally in 1966, dead, to be buried in the prairie above a clutch of Sandoz fruit trees.

Today, not all the section-line roads run through in the Sandhills, especially on a late-winter's day when sleet and what's left of the rotten winter's snow slicks the ruts in a hard shell of ice. One road, however, did run passably well from the blacktop to the gate of the old Sandoz place. From there I could park and walk. There was enough low haze and fog layered on the morning to give the land a ceiling, but it let enough sun through to spark the glass-clear ice that encased every single stalk of grass. In the snow, I could make out the grassless strip of a meander that must have been the driveway to the old place. Clearly, however, it had not been used all winter and was impassable. I set out afoot, sometimes on top of and sometimes breaking through the glassy crust on the snow, winding among the hills until I topped a bit of a rise that overlooked a long valley. At one end, maybe a half mile still farther up the road, stood a small, low farmhouse and a clutch of buildings, the homestead chopped in the hills by the stubbornness of Jules. At the other end of the valley, near the spot where I stood, there was a sign that said Mari's grave lay up the hill, above the orchard. Only there was no orchard, just windrows of dead trees.

The grave itself lies fenced, and there is a gate that swings to a counterweight that holds it shut against the wind. There is a simple headstone. On that morning of faultless quiet in the haze and ice, a couple of mule deer poked up from the grass and glided soundless up a nearby sand hill like worried spirits. Then I heard the first shot come booming across to me, across the valley, maybe from the house, I couldn't tell. I scanned the miles of open and could make out no sign of the source, no hunter, although one could have been concealed in the orchard that was still standing at the opposite end of the valley, about a mile away.

Then the silence settled into the place again, seeping in with the chill, damp air, and I tried to learn what I could from the yucca and the grass, especially from the few ice-cased stalks of prairie sandreed I gathered from the grave. Then another shot came, and this one spooked me. I left the place to its ghosts.

I walked up the snowbound road to my car, drove back to the blacktop, and headed north for a few miles until I spotted a sign tacked at the roadside that advised of a Mari Sandoz museum. This seemed safer than graves on a gray morning, so I took the turn and again found myself on a rutted-out meander of a two-track road, slick and treacherous through about four miles of the Sandhills. Along the way, there was an old schoolhouse but nothing else until the road arched around to meet a new wood-sided ranch-style house in a small forest of planted trees. This, besides being a museum, was the bird sanctuary and home of Caroline Sandoz Pifer.

Caroline herself, eighty-two, answers the door. She is the only person available to answer the door; she lives alone here. She tells me the museum is pretty much closed for the season, mostly because the tourists won't brave a Sandhills winter, let alone the museum's driveway. She says it doesn't matter that she's closed, I should come in anyhow. I tell her I don't mean to bother her and she says, "I wasn't doing a thing anyway, not a thing."

The basement of the house is the museum, and in it are the artifacts of Mari's life: photos, clothes, manuscripts, correspondence, rejection letters, and fan mail. Caroline is fourteen years younger than Mari; her sister was her teacher in grade school. Now she has care of all of this memorabilia and the literary estate. The latter is something of a bother, but she still manages such details as negotiating film deals, especially now that there is a fax machine at the sheriff's office in Gordon, a half-hour's drive to the north. The sheriff lets her use it when she needs to acknowledge Hollywood's existence.

I tell her of my visit to the grave that morning and about the gunshots that spooked me. She thinks this is pretty funny, because the gunshots are really a tape recording, a machine that periodically blasts away to chase grouse from the orchard so they don't eat the winter's buds. I am relieved to hear this, because the alternative is to believe in ghosts.

"Good," I say. "I thought the place looked deserted."

No, it is not deserted at all, she tells me. Caroline's older sister Flora still lives there. Alone. She is eighty-five. That mile or so of driveway to the only passable road looked unused because it is unused. Flora simply gets what she needs in the fall and stays put all winter.

What of that part of Flora's orchard near Mari's grave, the windrows of dead trees? Old Jules must have had a hand in establishing it, and it's too bad to see some history die, I say. Not really, says Caroline. "I tell you, this is not orchard country."

Caroline has her own little forest around her ranch, but it grew at some cost. The Chinese elm, they all died. So did almost everything else she has planted. "All these evergreens, they are all hand-planted. I bet we planted fifty trees for every one we got to grow."

Giving up planting, however, was unthinkable to Caroline, simply because she needed the trees to attract songbirds. There are dozens of species of birds native to grassland, ranging from raptors to burrowing owls, larks, and ground-nesting sparrows, but for reasons of their own, the generation of settlers that raised Caroline did not think of these as birds.

"We didn't have practically any birds until we got trees. I was probably twenty years old before I saw a robin," she says. "Now I've got tree sparrows and juncos. But no finches. They've got finches up at Gordon." There, the combined plantings of city dwellers have raised a bigger forest.

It would appear that an eighty-two-year-old woman alone in the world ought to live up in Gordon with the people and finches, to say nothing of Flora over on the next ranch, and so as tactfully as I can, I ask Caroline why she is still on the land and hasn't moved to town. She stares at me for a second, as if I had suggested she take up residence on Mars or as if the thought had never occurred to her. She says simply, "This is heaven."

And a big one. Caroline's land extends somewhat beyond the acre or so patch of trees round the house. It is a ranch of ten thousand acres, which she leases to cattlemen. She says that, to tell the truth, it's a bit small for her tastes, and she has been looking around to pick up some more land. The trouble is it's getting harder and harder to buy in lots as small as ten thousand acres, and she's not sure she can handle the more common lots of twenty thousand acres—thirty-one square miles—in which this land now sells. Single ranches hereabouts tend to be larger than Manhattan, and no one buys in smaller lots simply because it's not possible to make a living on such cramped quarters.

Many of Caroline's neighbors must supplement their incomes, even those who own ranches that would elsewhere be regarded as baronial. One man sells puppies. Another, says Caroline, makes leather belts and chaps and "sells 'em to the Indians." Another has combined nature photography with a sort of value-added manufacture. That is, he has discovered a market in the tourist trade for photographs of coyotes and has perfected methods of hiding that allow him to capture them in their element. Once he points and clicks, he shoots his model, skins it, and sells the hide.

Coyotes aside, though, the economy of the Sandhills has gone to cattle, not the intensive agriculture that a government once imagined, but the extensive grazing of the plains. Unlike more arid areas farther west, cattle fit better here. The Sandhills are not webbed with sensitive streams, so riparian areas do not everywhere provide a stark contrast to arid uplands, as they do closer to the Rockies. The cows don't bunch up and hammer the streamside shrubs and forbs. Because the water table is relatively shallow, the ranchers are able to pump water for the animals with small windmills scattered throughout the countryside. Given the large ranches, the cattle are more able to range from place to place as the bison once did, spreading the results of their grazing.

A few hours east of the Sandoz places lies another ranch, wrapped around the Niobrara River, this one the property of the Nature Conservancy, a national environmental group. It is a project for restoration ecology, an attempt to bring back a grassland the equal of conditions before European settlement. Its manager, Al Steuter, a plainsman and nationally respected student of grassland ecology, insists upon two things on his ranch: fire and grazing. The former gives him the most headaches in that the city folks who have bought vacation places along the scenic Niobrara complain of the smoke, but he burns his grass just the same, and he grazes it with bison and cattle. The plant communities here evolved under fire and grazing, this much is known. If they are to be resurrected, then both forces must be allowed a relatively free rein.

He says he is unable to buy enough bison to adequately stock and graze the preserve, so he falls back on cattle. "Cattle are as close a proxy for bison as there is." That's not to say the Sandhills grass cannot be overgrazed, and in fact the region has been from time to time, but a part of the fault of that, Steuter believes, lies not so much in cattle as in our patterns of land ownership. One hundred sixty acres or even a square-mile section is too small of a plot to graze cattle, even if their numbers are appropriately small. To survive on such a small spot, they must visit each location too frequently. Cows in some sense got a bad reputation because of this.

"We've scaled them inappropriately with our fences," says Steuter.

The land, however, is accreting now in larger chunks and this scale makes cattle fit better on the land. Such grazing may not be appropriate, say, in the short grass of the high plains, certainly not in the grasslands of the Great Basin, but this is the Sandhills, and the land calls forth grazing and has done so for twenty-five million years.

Attempts to call forth almost anything else have been a disaster, particularly all experiments with a plow. In Jules Sandoz's day the land was seen as an extension of the corn belt, the sand below be damned. The yeomen all took their 160 acres, sunk wells, bought plows, and ripped the bluestem and sandreed roots from the tenuous deposit of soil. What activates any ecosystem is its resources, its biomass, its zone of life. The sand that underlies everything here dictated a strategy for life: that it would slowly build itself from the minerals in the sand and concentrate its community in the sod. The sod was the key to the life of this place, its power.

"That's how the Europeans overcame all the power," says Steuter, "by turning it all upside down."

They beat the soil bare, but then the soil beat them. What slipped their notice was the elementary fact that they needed the power of the soil to derive their own. Drought was particularly hard on sand and farmers of sand. A series of dry years coincident with the depressed prices of the national financial panic of 1890 cut a wide swath out of the area's first wave of settlers. Corn that was selling for forty-eight cents a bushel in 1890 was eighteen cents in 1895 and thirteen cents in 1896. Banks closed, three of them in the small town of Rushville alone, just east of Gordon. Yeomen abandoned their land; some grass came back and with it some cows.

The Sandhills were a special impetus for some national legislation in 1904, the Kinckaid Act, which was a beginning of the recognition that Jefferson's 160 acres might not mesh exactly with what John Wesley Powell had learned about rainfall in the "Great American Desert." The Kinckaid Act quadrupled the homesteaders' allotment to a whole section, 640 acres.

Caroline Sandoz was born just after this. She was a part of the flush of prosperity that hit the hills with a new wave of settlement spawned by the Kinckaid Act.

"Until then, you got one hundred sixty acres and you couldn't even starve to death on that," says Caroline. Given 640 acres, though, the farmers "figured out how to starve to death just fine."

The Sandozes had a store and some gas pumps then, and Caroline remembers the period as simply "fun."

"It wasn't lonesome in those days. There was someone on every section. There were dances," she says.

Wrote Mari in *Love Song to the Plains*:

> From the first, there were diversions, perhaps a house raising of log or sod; church, Sunday-school picnics and prayer meetings and play parties instead of dances if the feet were what early settlers called "Methodist." Sorghum boilings, taffy pulls and walnutting came in fall, after the berrying and the plum, chokecherry and grape gatherings were over. Winters there were pie suppers, literaries with spelldowns and debates, shivarees for the newlyweds, and feather strippings where duck and goose ticks and pillows had to be eked out with chicken feathers, to be separated from their stiff and prickly quills by the young people at the party. When there were settlers in a community, political gatherings started, later perhaps protest meetings, and meetings to organize schools, churches and counties. Mutual protection committees were formed against this or that oppression, and these too offered some relief for the lonely, the isolated.

Almost none of this survived the waves of drought, the dust bowl years, and the Depression, and there no longer is someone on every section. The prairie sandreed has taken over the furrows and churchyards. This is not to say that the cattlemen prevailed over plowmen, a message borne by the lineage of Caroline's ranch. When Jules pursued his target practice and his other battles with the cowboys, his nemesis was the Spade Ranch, which finally went bankrupt in 1943, the year Caroline and her late husband bought it. The Spade today is Sandoz land, and on it the daughter of a plowman raises cows in a place she calls heaven, lonesome or not.

Down the road a few miles from her house stands the schoolhouse, officially Hinchly Elementary School, looking like a relict conjured by Laura Ingalls Wilder. Still, though, there are kids inside, three of them, or at least that stood as the total elementary school census, third through eighth grade, of the district in 1993; three kids from an area that could comfortably hold all of the boroughs of New York City.

Technically, it is a two-room school, but one of them is the living quarters of the teacher, Vickie Scarbrough. It's her first teaching job. The schoolroom has a Tandy computer, a TV and VCR, three desks, a copy machine, a telephone, and a rabbit. Scarbrough answers to a school board, a body of three people, consisting of the father of one of the students and the mother and father of another.

The students, Shad, Skyler, and Mike, regard me as every bit as crazy as Caroline did when I faintly suggest their lives here may lack certain qualities that kids find elsewhere. They have the rabbit and a basketball hoop outside. They have horses, Hereford and Angus cattle at home. They go to the library at Gordon every two weeks and on high holidays such as Valentine's Day a bus takes them to a similar school half a county away so that a combined student body might provide sufficient numbers for a proper swap of cards. To a man, the student body believes each will become a rancher when the time comes, and they look to the windows frequently, as if to suggest the time can't come soon enough.

Their teacher, meanwhile, concerns herself mostly with what goes on inside the building, the technicalities of education, bulletin boards, and libraries. She likes the plains well enough, or at least she says so, but has a hard time answering questions about her reactions to it. She's been here only a couple of months, dropped into this schoolhouse from a suburb of Dayton, Ohio, where there are convenience stores, malls, and a nearby McDonald's. Now she sits in a beat-up two-room building with a clean view shot at the horizon from the four windows aligned with the four cardinal directions. Questions about her reactions to this do not yet compute, as if she has not yet found the proper analog in her experience to give her the words to discuss it.

She says only this of the landscape: "They need more trees, green grass, and apple trees."

She will leave.

The Sandhills' work of regathering itself did not cease with the dust bowl years, but accelerated during the farm crisis of the 1980s. Here the hyper-industrialism of that period manifested itself as center-pivot irrigation. What the sand needed, went the reasoning, was a bit more water to call forth corn, water that was easily available from the shallow water table. Center pivots are long—maybe one-eighth-mile long—lengths of irrigation pipe mounted on tractor wheels and attached at one end to a well. One costs about forty thousand dollars. It pivots using the well as its center the way a school kid's compass uses its point to draw a circle. This one draws a circle of water.

Center pivots provide a new geometric variation to the old squared theme of the rectilinear survey. Now, when viewed from the air, whole areas of the plains are tiled with big circles like markers on a filled Bingo card.

Sand drains water efficiently, more efficiently than boosters of irrigation calculated, so in the late eighties one could buy center-pivot systems rather cheaply at bankruptcy sales in northern Nebraska. The small increase in production simply didn't offset the enormous costs of the systems. Farmers lost the land they had used as collateral to finance the idea. Caroline said several of her nephews went broke during the period, but especially those raising "little pigs," which meant they were raising corn. In the Midwest, hogs are considered the most efficient way to turn corn into money. The sand wouldn't hold the irrigation water, so it wouldn't hold corn.

The foreclosed land, as it did elsewhere, fell to the insurance companies and to the banks that held the paper on land and equipment. It gathered in larger bundles in corporate hands, but that was not the end of the story. The extensive agriculture that the grasslands demanded did not suit industrial agriculture's penchant for intensive farming, and the land rather quickly passed back to longtime sand-hills ranchers who had stuck to grazing and survived. That is to say, the once-plowed land called back the grass.

The long-term trend is clear. Take Cherry County, for example, the biggest of the Sandhills counties. In 1930, before the dust bowl era but after the Kinckaiders left, it still held 10,898 people. By 1960 it held 8,218, and in 1990 only 6,307, a decline of 42 percent in sixty years. The historian Frederick Jackson Turner's famous formulation for the American frontier defined it as that place with less than two people per square mile, and on that basis, he and the U.S. Census Bureau declared the frontier dead in 1890. In 1990 Cherry County held 1.1 persons per square mile.

Of the fifteen counties that more or less make up the Sandhills, all declined in population between 1930 and 1960 and all declined again between 1960 and 1990. During the sixty-year period, the population of the fifteen counties dropped by 44 percent. Arthur County held 462 people in 1990, six tenths of a person per square mile.

Blocks of land no longer come in the ten-thousand-acre pieces that Caroline Sandoz wishes to buy, but sell in chunks on the order of ninety sections, more than fifty thousand acres. The people buying them are young people of the Sandhills, and they are ranchers, says Caroline. It's all fine with her. True, the country is more lonesome, but she's learned to live with that. She has her birds, but she has noticed something new about this latest generation of ranchers.

"Nobody ever plants a tree. Wouldn't you think someone would come along and plant trees?" she says.

• •

When Al Steuter, the manager of the environmentalists' ranch, makes a move, he does so slowly. He is trying to run a place that is first and foremost an experiment in maintaining biodiversity, but to do so, he takes his first examples from the way the ranches around him operate. This is not the contradiction it would first seem. He is paying attention to the history of the Sandhills and behaving first as a biologist who takes natural selection seriously. The place—gentle, rolling, and soft brown—can be as unforgiving as the wind and sand. This land has sent many people away or to bankruptcy or to early graves dug by their stubbornness and prejudice toward the nature of the land. Those who have survived have done so by learning something from the place. They did not reform the land; it has reformed them.

The power of the Sandhills was lost to plows and fences, but now it calls forth grass and big places, big open places, and a few people who will listen to this and survive. The ranching culture of the West may seem to some bigoted, harsh, xenophobic, and mindlessly conservative, and in many ways it is. But in many ways the rest of us are in the same position as Steuter. How will we make it to a future that respects nature without learning what the surviving ranchers have learned, especially those who have weathered the hard lessons of blowouts and sand?

David Hansen, an organic farmer and a pastor in three churches on the south edge of the hills, says that providing a church's cohesiveness to a scattered and independent flock has required some adjustment. For instance, on Sundays in early spring, Hansen used to lose his congregation to branding season as the families spread across the hills in communal work bees, branding their neighbors' cattle. Now the church has a spring branding service. Each family brings its branding iron and burns its marks into a board, which becomes a symbol of community.

It is, though, a community that holds not so much because of institutions and symbols but flows from an attachment to place.

"We're here because this is where we want to live," says Hansen.

The solitude of the prairie is like no other, the feeling of being hidden and alone in a grassland as open as the sea. Walking toward the horizon through the hills, tawny and loose like the folds in a cougar's skin, one has a sense that over the next ridge there will rise a brown cloud of bison and over the next, the Pleistocene, unspoiled. Unless one has walked pure prairie, it is difficult to imagine how such a sense of freedom can flow from a landscape that is the giver of harsh rules.

Hansen says his favorite sermons deal with "concepts of God as the one in charge. So many things are beyond our power, and therefore we live within restrictions."

He can be forgiven for preaching this message to the converted.

In a café just outside of Valentine, Nebraska, a ranch family gathers for an evening meal at the table next to mine. Seated at the head of the group is the father in a big gray Stetson that stays on his head at table and a crisp snap-button shirt. The table has a rim of kids in smaller hats. Nothing much is said, but it is not the hostile silence of family dinner that many of us know. It seems only a family with a habit of quiet and deference to one another. There are a few soft-spoken questions. Siblings seem interested in the answers, but most of what needs to be said is handled with a few words and eyes. The man, with close-cut hair, clean shaven, probably forty but looking thirty, is eating what I suspect is the world's best steak and washing it down with what I know to be the world's worst coffee. All of this scene flows as remarkably and unremarkably as a sunrise.

There are a million questions the reporter in me wants to ask him. I can't say whether he knows the answers, but I suspect he does. He knows things the rest of us need to know, certainly things I do not know as an outsider. But I have been visiting the Sandhills long enough by now to have realized a small fact: What he knows cannot be said to an outsider, even if I were smart enough to phrase the right questions.

11

Seeds

There is not a single scene that can show the evolving ethic of a 130,000-acre ranch. Two hundred square miles affords no view of the whole, particularly when the land rolls out in intermontane valleys toned by the shifts and shades of the northern Rockies. Particularly when the human players of the place run the gamut from hard-set cowmen cut of a piece from Montana to media moguls and megastars. Particularly when these valleys that are at the epicenter of cow country contain not a single cow.

Still, the best telling of the place may be a pile of junk at its center, New Age junk the land seems to have shrugged off for the cowboys to haul away, mostly piles of pulled wooden posts and coils of wire that once stood service as the miles of cross-fencing that webbed the Flying D. The perimeter is still fenced, but in the big heart of this place the land flows free and unfenced out of the Spanish Peaks just northwest of Yellowstone National Park, across the creeks that drain them, across the big waters of Madison and Gallatin—Missouri headwaters—on still until the land butts against a tourist highway that threads Winnebagos from Bozeman south to the park.

It is a mark of the place's evolution that Bud Griffith, the ranch's manager, points to the junk pile with some pride. What Bud says about the Flying D counts because it is his ranch more than it is the on-paper owner's. Bud came here first in 1959, then a high school dropout fleeing Vermont, and started cowboying in the steel-wire and hard-mouthed-bronc days. His first job was on the Flying D. By 1961, he was manager of Spanish Creek Ranch, working for Mr. Jim Ray. Spanish Creek was a broken-off piece of the Flying D, but now the two have been put back together.

While Bud was at Spanish Creek, a California company owned the adjacent Flying D and leased it to cattlemen, who gave a good goddamn about nothing but pounds on the hoof; they beat it nearly to death. In 1971, Mr. Jim Ray sold the Spanish Creek to Mr. Robert Shelton, an heir to the famous King Ranch out of Texas, and Mr. Shelton knew a lot about grass. Bud Griffith stayed on to run the place. In 1978, Shelton saw a chance to buy the Flying D. He did and reassembled an empire set in cows, a modern ranch tightly managed for beef, with some irrigated crops and hay meadows in the lowlands, some elk for the dudes to shoot in the aspen-stemmed draws of the high country, and a closely managed cattle range in between.

Bud raised his two boys, a string of saddle horses, and he ranched, a life he considered bliss, but then new owners bought the place in 1989, and life went weird. Bud says his first thought was simply to pull up and leave. The new owner had some odd ideas, and when a man is rich enough to drop out of the sky and write a check for 110,000 deeded acres, 20,000 more leased, there is every reason to believe his odd ideas will prevail. Especially if the new boss is, as Bud tells it, stubborn, opinionated, and out to prove a point to the world.

"I could stand to see the cows go, that was bad enough, but the horses," says Bud. "My first thought was to hell with it."

He says what worried him the most was the prospect of never again being able to face any of his neighbors, cowmen who would consider his new way of working an act of betrayal.

"I had heard everything that was ever said bad about him before, being a staunch cowboy," he says. "I hated to see the cows loaded up, but then I got to thinkin': The land's still here. Besides, Ted and Jane's been nothing but good to my wife and me since they came here."

As Bud is recounting all of this, the two-way radio in his new red Chevy pickup comes to life and the emperor of cable television is on the air requesting from all available ranch hands a report on the fishing conditions at Cherry Creek. Cherry Creek is a trout stream and the new boss is a fly fisherman. Bud grins as another two-way somewhere on the ranch files the story to the ranch's anchor desk. Then comes on the radio an exchange among hands and foremen as to how best to ferry Jane Fonda's lunch and its special-order loaf of bread to where she is fishing that day on Ted Turner's other ranch, 70 miles away.

Above this fray, Bud widens his grin until it reads, "Ain't life peculiar," and we drive on, the truck side-slipping a bit in the late spring mud at the ranch's high, wet end. We roll up near the ranch's boundary with U.S. Forest Service land in hills of quaking aspen (call them "quakers" up here) and pine. Hundreds of elk graze a draw, regal adults tending calves of spring. We go over another ridge, past a black bear rooting for shoots on a side hill, past the eared triangle of a coyote's head spying and slinking behind a stump, past eagle, blue grouse, and meadowlark; then the Chevy breaks a ridge top that reveals the possibilities of the planet.

Before us lies a broad valley that is a deep green sweep of grass. There is a slick little creek set in willows. The backdrop rim is the Spanish Peaks, a range of the Rockies still shouldering early summer snow.

"The country's got her suit clothes on this time of year," says Bud. Below our ridge rippling in brown waves across the valley are several thousand bison. The Flying D's new cash crop is buffalo. We drive among the chocolate-brown cows and bulls, the yellow-brown newborn calves, thousands of them.

All the time Bud is reading hair, a cowman's trick, a way of assessing the animals' general health. Some of the animals are late in shedding last winter's pelage, a fault that correlates with the rangy and slacker-looking animals. Late shedding means they are sick. These are culls and a minority. Most of the bison here are magnificent. The bulls, just these past few days joining the cows for the annual rut, are round and tumescent, glary-eyed, maybe seeing the Chevy as possible competition. They barely move out of our path.

Bud checks the mineral box, a mix of salts and the like devised by bovine nutritionists, about all there is to managing a herd of bison in spring. The mineral use is his window to their well-being. Any break in their routine signals trouble. They eat a lot of mineral in spring when the grass is growing and then slack off later. Sometimes they wander to a hillside striped by a band of clay and dig their own minerals.

Bud's been watching the bison for five years and if there has been a single important aspect shaping his learning curve, it has been the lesson that bison are best managed if they manage themselves. Nutrition and feeding are mostly a matter of giving the animals the run of the ranch and letting them take what they need. Herding is a matter of watching them until they decide to go where Bud wants them to go, then playing along.

The animals have been in this very landscape for at least four thousand years with no human help, so the better part of wisdom is staying out of their way. Their evolution has taught them the land better than we will ever know it. Unlike cattle, they can survive most winters without any supplemental feeding. The ranch used to feed hay to its cattle seven months out of the year, but now feeds almost no hay to its bison. Managing three thousand cattle would require, conservatively, a dozen cowboys. Bud and two other people manage three thousand bison.

The cattlemen, politicians, and editorialists of Montana have not yet tired of flailing at Turner. Among the ranchers, Fonda is still referred to almost exclusively as "Hanoi Jane." The state's political establishment says Turner's ownership threatens the cow culture of Montana. They say the Flying D has become nothing but a rich man's toy, a charge that Turner denies. He says he's in it for the money. There's a market for bison, both for building herds and for slaughter. They fetch about twice what a cow does, pound for pound. Because of their compatibility with the land, they cost about half as much to raise. Turner believes his Flying D is the economic future of the West. The future it may be, but this is about more than money.

On our tour, Bud seems to display the most pride when he shows me his bison, but almost as much when he shows me his weeds, or more precisely, increasing lack thereof. Along the steep draws near the Madison River, the Eurasian exotic leafy spurge had more or less taken the landscape. Unchecked, it could do here what spurge does everywhere. Turner has experimented with all available controls, even making his ranch something of a research arm of Montana State University, but in the end he has found what many of us have found, that the only way to deal with this threat is the herbicide picloram, applied at rates that cost as much as one hundred dollars an acre.

After years of experimenting with combinations of chemicals, grazing, and fire, the Turner ranch is becoming close to spurge free. The achievement was not cheap and likely wiped out any pretense the ranch made of showing a short-term profit. This is the most sobering of lessons for Bud, a demonstration of a conflict of the heritage of ranching. He is loyal to the human culture of the place and would hate to see it go, but understands this: The sufferings of the landscape in the West derive from the necessity of profit in cow equations. Overgrazed land is susceptible to invasions by weeds; even moderately grazed land is.

Bud says there is no way his neighbors, even well-meaning ones, can properly deal with the injuries to the land and still get by economically. You can't spend a hundred dollars an acre to control weeds and still show a profit, and above all, ordinary ranchers must profit to survive. Proper care of the landscape is the prerogative of rich men like Ted Turner who can afford herbicide, can afford to leave hundreds of acres of grain fields unharvested for ducks, geese, and pheasants, can afford to allow ground squirrel populations to explode until hawks, black-footed ferrets, and owls resettle the equation, can afford to give grass to three thousand elk and to open his gates to archaeological teams from nearby universities and reservations.

In the long run, all of this may produce a profit, in the narrow economic sense, but the notion of profit just now on the Flying D runs broader. Just now, one can measure it best in grass, the waist-high layers of native bunchgrasses decked out in full community of shrubs and flowers swelling to cover the miles, all grazed. It can be measured in streams that run clean in tunnels of vegetation, a rare sight in the overgrazed West. The Turner ranch is as healthy a grassland as I have seen anywhere in the West, including in national parks.

Such scenes can be duplicated on land grazed by cattle, but only if those cows are intensively managed. For instance, to prevent them from trashing lush streamside vegetation, cattle would have to be fenced away from the streams. Given the run of the place, the bison still stomp a stream bank now and again, but they do almost no damage. They are a part of the place, a part of the healing, not the injury.

It was Turner's idea to pull out all the cross fences on the Flying D, and Bud says he fought it hard at first. Those miles of wire were his way of handling his cows. He had been taught to be a responsible rancher and to do that he needed cross-fencing, separately fenced pastures so that the cowboys could force the animals into rotating the areas they grazed. Cross-fencing is a way of ensuring that the cattle don't visit a given area too frequently. Forcing behavior was the first way Bud learned to handle animals. He learned to break cattle as one broke horses. "We used to ride them pretty tough and broncy," he says.

The fences came down, however, and Bud didn't quit the Turner ranch. Instead, he came to see his new job as a chance to relearn animals, which eventually came to be more inviting than simply herding cows. He learned a lot the hard way, but slowly it came that the key to herding bison was not so much herding them as it was following.

By then in his fifties, Bud suddenly cut a new deal with his stock. He does not own them. Instead, he knows them, and through knowing them he knows the land. In all of this, he has come to agree with his boss on certain matters, especially that the results are unarguably better for the land, but he is still enough of a cowman to believe cows have been unfairly accused. Like the land, they are still full of our fences.

"Cows are what we made them," he says.

We've herded them, moved them, and bred them, crossbred and hybridized them until our wishes and ignorance are genetically encoded in them as stupidity. Their behavior is as mechanistic as man's and they've unlearned the habits of the land. The bison are still wild, their genetic coding and instincts still intact. They are social, headstrong, and hardy.

"I've got a strong feeling that when you take the horns off an animal or castrate a bull, you have a big effect on his social life," Bud says, as he shows me a sharp-banked knoll on one edge of the ranch.

What's odd about this short hill is that for some reason or other the buffalo like it to the point of spending entire days walking up its slack side to the peak, then gamboling, rolling, and frolicking off the steep side, all to no apparent end other than to do it like kids on a sled hill. The bison, says Bud, are simply playing, and he's never seen cows do that. Once in a while, a heifer will kick up its heels, but cattle don't spend whole days in organized play. This is the best clue Bud has of being on the right track.

In October of 1993, trucks threaded the high plains highways bearing three hundred bison home. Their destination was five thousand acres of restored grassland, tall-grass prairie, near the northern

Oklahoma town of Pawhuska. A full-blown media entourage and one thousand spectators greeted them. The Osage Indians from roundabout held a ceremony. A western swing band played and there was a black-tie-and-jeans dinner. General Norman Schwarzkopf was there and was field-promoted to honorary Osage chief by some members of that nation.

"We are basically putting the train back on the tracks and restarting the engine," Bob Hamilton, the manager of this project, told *The New York Times*. Hamilton works for the national environmental group the Nature Conservancy, and the project is the Tallgrass Prairie Preserve, 36,600 acres that eventually will hold 1,800 bison. The project has been featured in beer company advertisements in national magazines. The conservancy is using the majesty of the bison to fire some national imagination about the majesty of grass, so at bottom the project is about plants. Saying the bison are back is only another way of saying the prairie is back. Although not as intensely photographed or advertised, the jointed goatgrass, rattlesnake fern, spleenwort, and lady's tresses also have returned.

All of the grassland West is a candidate for restoration ecology, a candidate for resurrection, and that resurrection must include bison. The prairie between the Rockies and the Mississippi River was held by three pillars: bison, fire, and grass, and the place cannot live again until all three return. Nationally, the Nature Conservancy has been a leader in recognizing this. The preserve in Oklahoma is only one of its efforts at putting the world right again. This is not just another pen of bison kept as a roadside attraction.

The conservancy's involvement, though, is in some ways not that much different from a range of efforts to preserve the bison carried out through all the twentieth century. And in a way it is different, a shading into a new phase. Efforts at preservation in the early part of the century focused on preserving just the animal, which means preserving the animal out of context. What the conservancy is doing now is resurrecting the context, the system of which bison are a manifestation. The bison that exist now are in a real sense the seed that has been preserved. Now we use that seed to regrow the prairie.

But the conservancy's efforts are similar to past efforts because of its notion of a preserve, a place separated from the main of economic life. As such, it is a direct descendant of the conservation of Theodore Roosevelt. It is in line with national parks and wildlife refuges, a recognition that our way of life is sufficiently brutal as to force us to remove some of nature from it. That's the defensive action. Now, however, we need to understand that the hope for both the animals and for ourselves rests on integrating nature into our economic system. To some extent, the conservancy is beginning to explore this idea, but to a much larger extent that's exactly what places like Turner's ranch are about. Turner's ranch is the promise, but it also signals the problems ahead.

The problem is not that the coming battle lines are without precedent. They are as old as civilization itself, aligning perfectly with the split between Cain and Abel. The problem is more this: Given our culture's alienation from nature, we are likely to confuse the issues.

By any real measure the goal of preserving the bison was a success. Working from that core group of eighteen animals at Yellowstone National Park in 1872, along with small groups in zoos, preservationists have built the public herds to close to twenty thousand animals in Canada and the United States. Of these, the largest groups are the ones held by the U.S. federal government, a little more than five thousand animals, with close to half of those still at Yellowstone Park. There also are federal herds in Montana, North Dakota, Nebraska, South Dakota, Wyoming, Oklahoma, and even Kentucky, largely living on the refuges of the U.S. Fish and Wildlife Service.

Independently, nine states' governments, those of Alaska, Arizona, Idaho, Kansas, Minnesota, Nebraska, South Dakota, Utah, and Wyoming, maintain herds totaling another four thousand animals. There are four herds of plains bison in Canada, two in Alberta, and one each in Manitoba and Saskatchewan. Canada also has the only surviving population of what is generally regarded as a subspecies, the wood bison, so named because it inhabits the fringes of subarctic forest on the north rim of the grasslands. There are close to three thousand wood bison in seven herds in Alberta, Manitoba, the Northwest Territory, and the Yukon Territory. Canada also has another three and a half thousand animals that are a hybrid of the plains and wood bison.

The public herds, however, are completely dwarfed by those in private hands, mostly buffalo ranches like Turner's. The total population of bison—public and private—stands at about 150,000, leaving more than 130,000 animals in private hands. The total number is a long way from the fifty million that once inhabited the plains, but a long way, too, from the three hundred that existed at the turn of the century. And while it can be argued that the federal preservation effort saved the bison from the fate shared by so many plains plants and animals, clearly it has its limits. Buffalo ranching is taking over and boosting the number, for the simple reason Turner points out, that a bison costs half as much to raise and sells for twice as much as a cow.

There are bison ranches now in all fifty states, but the big herds live in the West. A single ranch in Pierre, South Dakota, has more than three thousand head. A subset of the ranching movement is building among a consortium of Indian tribes throughout the West, who have begun returning bison to tribal lands. Late in 1993, that group began a cooperative effort to process and market hides from the animals slaughtered. The leather will be pushed back into other tribal enterprises, especially traditional leatherwork and crafts. Painted bison skulls sell for about three hundred dollars.

I could not see exactly what the old men were doing, nor could I have understood what they said, even if I knew the Blackfeet language. It was a mumbled chant, and they were elders, so all of the few hundred of us there that day kept a respectful distance from the cattle chute where the old men huddled. There was some business with trade blankets and the burning of some smudge, that much I could see, but the rest was the affair of the elders and nobody else.

The aluminum frame of a cattle trailer was backed to the chute; the diesel truck pulling it sat idling in the chill prairie air. The driver, a kid from South Dakota, was taking it easy, taking in the end of what had been a wild day, now at its somewhat anticlimactic climax, which the elders were blessing. Inside the cattle truck, the cargo—half a herd of eighty-four yearling bison—kicked and flailed at the aluminum grate. They did not yet know they were going home.

Around the weathered wood corrals that would receive the bison, half the Blood tribe gathered, awaiting this homecoming. A few miles out across an unbroken reach of last fall's grass, out on the blacktop highway, the rest of the tribe waited, angry. Just hours before, some of them had lain on the highway, trying to block the return of the bison with their own bodies, and they did until the elders broke it up. The Mounties had come, and the homecoming turned ugly, but in the end, Harley Frank had won.

Harley is the tribe's young chief, maybe forty, clean-cut, and sharply dressed in jeans and a nylon jacket over which he wears the Victoria medal, the tribe's symbol of sovereignty in Canada. He had needed to invoke all of the power he could muster that day, diametrically opposed as he was to the tribal council, the protesters, and even the Mounties. He had shown the medal out on the highway in a standoff with the police. The Mounties had wanted to keep the peace by having Frank back down and order the trucks off the reserve. The Mounties did not want to confront the protesters who were battling Frank.

"These animals and these trucks are coming across. These animals are the spirit and the will of the people. . . . I'm ordering you to move [the protesters]," said Frank, shoving the medal into a

Mountie's face. "You see this medal. This is Queen Victoria's medal."

The Mounties stood, stuck between the two factions of the tribe. The protesters—apparently traditionalists because of their dress and talk, but it's never that simple—laid themselves across the highway near a bridge at the town of Standoff, in Canada's largest Indian reserve. It was indeed a standoff, and here is how it ended: In between the protesters and the Mounties marched a bunch of little old women wearing trade blankets, some of the cragged faces clutching cigarettes, all of them clutching wrath. They ignored the Mounties and headed straight for the prone men. The women, one of them claiming to be 107 years old, simply lifted the protesters, carried them across the road, and flung them into a ditch. Then the trucks eased on.

I had first encountered this group of women, members of the Buffalo Woman Society, earlier that morning, a couple of hours south of the reserve at the border town of Coutts, the port of entry from Montana into Alberta. There, in sight of Montana's Sweet Grass Hills, Frank, the elders, singers, and dancers had met the two trucks from South Dakota. There had been a parade. Oblivious to the ice on the March morning's wind, the assembled Blood came drumming and singing the animals across the border, the old women in the lead.

The women in bright blankets, the drumming, and the television crews had caused a bit of commotion at the border checkpoint, delaying the stream of trucking trade. Some of the truckers found out what was going on and hauled out video cameras of their own, becoming part of the event. One trucker came into the border station out of the cold bite of the wind: "What's going on?" He was told. "I'll be doggone. I'll be doggone."

Others grumbled, and I heard one say: "Yeah, if you got a look at the tag on one of those blankets, it would probably say 'Made in Taiwan.' "

Maybe, and it wouldn't have been at all inappropriate for these Blackfoot (the Blood are a subset of the Blackfoot settled in Canada) people to wear imported blankets. They have been trading for their blankets for several hundred years, long before free trade agreements made Coutts a bureaucratic island in a river of aluminum trucks.

Bringing in the bison from the U.S. was Harley Frank's idea, the central notion of his agenda for his people. Although the Blood nation is deeply in debt, he pried loose the $100,000 in tribal funds to buy the bison from among the animals used to make the film *Dances with Wolves*. Frank, though, is not a traditionalist—that is, not an Indian given to returning to his people's past. He took his advice from and trained his buffalo keepers on the Turner ranch. This new herd of young cows is to be the beginnings of the tribe's economic development.

The reserve is poor, but it occupies a gorgeous piece of prairie stretching against the Rocky Mountains on the clean, high plains. Frank believes the animals will become the core of a tourist trade, and eventually the core of livestock trade. They'll produce leather and meat, in the spirit of the Nebraskan who photographed, then shot coyotes.

There is nothing new in this. John James Audubon shot nearly all the subjects of his famous drawings and paintings. The plains make a practical people.

"We have been on welfare. We have been on rations for too long," says Doreen Rabbit, a supporter of Frank's plan and the tribe's cultural director. "Finally we will be able to provide for ourselves."

The caravan had not made it far north of Coutts on the lonesome swing of a highway when the Mounties pulled it over, found Frank, and warned him of what was ahead at the reserve. The dissidents had gathered at the bridge near Standoff. Already word of the protest had spread. The younger schoolchildren of the Blood were to be let out of class that day to celebrate the return of the bison, but the holiday was canceled to avoid the brewing trouble.

Some of us would like to think the churn of events since the founding of the American Indian Movement and the rise of traditionalists have given us a new way to read Indian reservation politics that all issues boil down to confrontations between radicals and assimilationists. Certainly, reservation politics do boil, but the split is not that simple, and it was not that day on the bridge.

It may have looked that way as the miles-long line of pickup trucks and stock trucks made for the bridge to find a vocal blockade of young people, looking like the radical young people in places like Wounded Knee, South Dakota, in the seventies. The protesters all wore braids and bore hand-painted signs and belligerence. This attitude contrasted with the profile of the clean-cut, quiet, and conservative Frank, but allied with Frank stood the tribal elders, most of them "speakers," meaning they spoke only their native language. The confrontation at the bridge played out almost completely in the Blackfeet language.

After the women broke the blockade, I located a man in a pickup truck who was one of the leaders of the protesters. As determined and as well spoken as Frank, he invited me to sit down. Dennis First Rider disputed Frank's estimate of the cost of the animals, placing it closer to $400,000, a major sum for a tribe already $3 million in debt. He said he had nothing really against the buffalo, and in more secure times, the animals would be welcome. It was just that the tribe had other needs to provide for its future, and money would be better spent modernizing agricultural facilities, such as irrigation works. This was not a battle between reactionary and progressive. It was a battle over who gets to define progress.

"One thing we don't need is irrigation," one of Frank's supporters had shouted at the bridge. "We need the buffaloes. These are for our children."

After the confrontation at the bridge, the stock trucks moved straight up the highway another mile, then left it to bounce across the prairie toward the horizon. A few pickups and a van full of elders followed, but the Mounties blocked the rest of us, figuring to head off further confrontation. We simply cut back down the highway a mile, then drove out across the prairie as well, a herd of pickups and Jeeps stampeding toward the climax of this story. I was able to drive within maybe a mile of the point where the stock trucks had disappeared in a draw. I parked and began hoofing along with all the rest.

An Audi 5000 bearing four Blood men pulled by and stopped. One rolled down the window and said if I wanted a ride I should climb on the trunk, and I did. They later introduced me to some old men gathered at a tipi, Charlie Fox and Jim Shoots Both Sides, Frank's predecessor as chief. They were wearing bright red shirts, Stetson hats, and treaty jackets, the very same blue wool jackets their ancestors had worn to sign the tribe's treaty with the envoys of Victoria. These old men said they were all glad to see the buffalo come, then left me so they could take a special position in the final few yards of the procession when the first of the two trucks backed its way to the holding corral. A horseman in a shiny nylon Chicago Bulls jacket rode point in the buffalo's last few feet to home.

Other old men, the spiritual leaders, gathered in that chute with sweetgrass and sage smudge, and I crowded right next to it to catch some of the smell and the ring of native song.

Frank spoke: "Let us tell our grandchildren that on this day the power of the Bloods returned. We have become enlightened to the coming age of our people."

Then the bison were freed from the trucks.

In August of 1993, the International Bison Conference and Trade Show convened on the far eastern edge of the grassland at La Crosse, Wisconsin. It was a chance for six hundred private bison ranchers to further their cause and trade, but the bulk of the proceedings was buried in press reports by an account of a featured speech. Both Ted Turner and Jane Fonda attended, but Fonda's appearance drew protesters.

The longtime liberal activist was there to spur on the bison producers. Her logic and Ted's is simple and correct. They wish to preserve the life of the land, and doing so in much of the grassland means bringing back the bison. This, in turn, means bringing the bison back into the livelihood of the people.

"Because of Ted, I am part of the industry and proud to be. . . . We love the buffalo, and we want to encourage and build our industry."

Like Fonda, the protesters were liberals, to the extent that label means anything. In fact, that was the protesters' point. They were there to assert that Fonda's support of the slaughter of bison was a selling-out of her liberal principles. "We are appalled Jane Fonda would be a part of this," said Tracy Reiman of Washington, D.C., one of the protesters. Reiman was representing People for the Ethical Treatment of Animals.

As it was in Alberta, the protest in La Crosse rested on conflicting definitions of a responsible future. Unlike the protest in Alberta, however, this one more clearly splits on some age-old lines. The animal-rights movement is urban and derives from people who follow civilization's idea of progress as it is removed from nature. In their epithets aimed at Fonda, we can hear an ancient accusation, the same the Chinese leveled at the Mongol nomad and the same the Jeffersonian yeomen leveled at the Kiowa, Cheyenne, and Sioux. We hear the epithet: "Barbarian."

Rights for animals. Ethics for animals. Rights and ethics, like literature and law, are ideas derived from the plow. We need them to live successfully in civil human communities, but nature confers no rights. Nature confers life.

Why is it not ethical to kill and eat a single bison? A single bison does not stand alone, is not an individual. It is, rather, a manifestation of a place, the net result, the capstone of fire, wind, and grass—grass to the horizon—and of the hundreds of plants that live in it, and of the fungi, insects, the birds, the wolves, the prairie dogs, ferrets, burrowing owls, compass plant, horned lark, sunflower, and cone flower, all of these things, and can only be understood as such. The return of our eating bison marks the return of all of these things to our lives.

Why is it unethical to kill and eat a bison when all the rest of the bison and all the prairie life they stand for will go on? Why is it ethical, in the name of rights, to save a few bison in parks and zoos and eat instead wheat, to turn loose the plow that ensures, above all else, that nothing goes on? Why is the plowman not the barbarian simply because no one sees the blood on his hands? Where is the logic here? I mean to assert there is none, but by this assertion, I do not mean simply that the animal rights people, with whom I disagree, are illogical. That is civilization's trap.

Writing of the Inuit hunters of the Arctic, the Danish novelist Peter Hoeg said: "Compassion is not a virtue in the Arctic. It amounts to a kind of insensitivity; a lack of feeling for the animals, the environment, and the nature of necessity."

It's the same in the plains.

The "ethic" that civilization would impose on the land is as artificially derived as the chemical fertilizers it would impose on a cornfield. Aldo Leopold began tackling this notion a couple of generations ago with a call for a land ethic, which we took to mean an exhortation for an ethical treatment of the land. This has been the impetus for conservation.

But I think he meant to call for something deeper: an ethic derived from the land. Harley Frank had it right to assert that the return of the bison marked the return of the power of his people. Power, when it derives from the land, is a land ethic. Real power and real land, but perhaps we can understand this best not as land or bison or people. Perhaps we can understand it as each of these is the manifestation of each other, and this we understand as seeds.

I like to think of Pauline Drobney as perhaps a granddaughter of Willa Cather's character Ántonia Shimerda. She is a native Iowan, raised in a flat place, a packed, black earth sea annually roiled by waves of corn. She has a fine-cut face, intense, slightly fierce, with crisp, straightforward eyes. Unlike most botanists, she talks to people, mostly a rapid-fire stream of the details of her working life, which is important in that her life is eight thousand acres of badly beaten land. She began to work for the U.S. Fish and Wildlife Service only recently, and unlike most federal bureaucrats she will tell you everything she knows, or at least as much as she can in a couple of days' time, the considerable catalog of facts along with her doubts and fears, blessed as she is to practice her science in a time when the doubts, hopes, and fears of a scientist are admissible. One senses that all of this openness comes to Pauline by her nature, but it is all the better because openness will be demanded by her job. Some of our people are beginning to understand that the culture of plants is the same as the culture of people. Pauline understands it better than most.

"It's only when the people of the landscape value and cherish their ecosystem that there is a possibility of restoration," she says. "This is very much a search for our place in the landscape."

"This" is her project, officially the Walnut Creek Preserve, the latest piece of work by the Fish and Wildlife Service. In a particularly satisfying manner, the project just outside of Des Moines in the easy rolling hills of southern Iowa is a legacy of nuclear accidents at Chernobyl and Three Mile Island. The federal government probably would neither have considered nor braved the thorny politics of acquiring a couple of square miles of prime cornfield real estate—tiled, contoured, fenced, and cross-fenced—simply to return it to grass. Our people, as they are now constituted, would not regard this a proper use of prime agricultural land.

We would, however, allow those same cornfields to become the site of a nuclear power plant. Redlands, Inc., acquired the original 3,500-acre site in the early seventies to develop the plant, but the political fallout from the nuclear industry's two biggest black eyes sunk the plan. Pushed by Representative Neal Smith of Iowa, the federal government began acquiring the site in 1991. Then came additional money to buy more land. As of 1993, the site had grown to 5,000 acres, and Congress has authorized it to expand to 8,500 acres, bought from willing sellers as adjacent farms go on the market.

The plan is simple in concept: to restore all of this land, the cornfields, the wood lots, even the roads, and hog barns, to native, tall-grass prairie. This has never been done before. There are preserves all around the plains, but all of them are relict sites or, at worst, sites that have been overgrazed, rested, and restored. So far, no one has taken this much farmland, plow land, and put it all back to grass—head-high grass, big bluestem, Indian grass, and a full complement of forbs like leadplant, Maximilian's sunflower, coneflower, compass plant, gentian, and orchids, and, in the end, if all goes well, the capstone to the plants, bison and elk.

At the outset, the nature of the project dictates a thread of logic that weaves it back into the surrounding community, but this implies a real and significant shift in the philosophy of the Fish and Wildlife Service in particular and the federal government in general. This departure reflects an understanding of biodiversity. Until recently a buzzword among bureaucrats, a public-relations tool, biodiversity has become, among the best of our scientists, a concept that is a serious and necessary axiom.

Heretofore, the Fish and Wildlife Service has been in the refuge business, but most of its preserves have amounted to little more than duck farms. The public wants ducks to shoot, so it became the mission of the service to pump up the land to produce ducks. The service was not above, for instance, planting agricultural crops to feed ducks and exterminating natural predators, even other birds, that cut the "harvest." There was no notion of balance, nor was there a corresponding sense of protecting the integrity of the plant communities of a given refuge.

In the same vein, the U.S. Forest Service exists to produce trees; the Bureau of Land Management, to produce grass for cows; and the National Park Service, to produce scenery and rubber tomahawk stores for tourists. Slowly, though, a new ethic is percolating from the ground up into each of these agencies.

Under the old ethic, the Fish and Wildlife Service would attack a project like Walnut Creek simply by seeding the cornfields to grass, maybe even exotics, then bring on the bison and call it good. Under the new ethic, the role is to first understand and, to the extent necessary or possible, nurture the entire range of life that inhabited a place before industrialism wiped it out. Walnut Creek is self-consciously the agency's first biodiversity preserve. Pauline Drobney, formerly a freelance botanist and a firmly rooted conservation biologist, hired on as its prime scientific spark. She overcame a long-standing and freely admitted antipathy to the feds to take the job.

"I'm a native Iowan. I had job offers all over, but I made a conscious choice to come here. We plowed up something here, but now there is a possibility for restoration."

Iowa is all possibility. Its surface area, in terms of native vegetation, is the most disturbed of any state in the union. Once covered with tall-grass prairie and a few fingers of oak savannahs, the state now holds less than one half of one percent of its original habitat. By comparison, the amount of old-growth forest remaining in the Pacific Northwest—the region where our most vociferous environmental battles rage—makes that region positively pristine.

Most of Iowa's remaining tall grass exists as what the locals call "postage stamp prairies," relicts along railroad beds or cemeteries or a farmer's inaccessible field. Other than Walnut Creek, the largest example of native communities extant is about two hundred acres.

These relicts are the first hook for Drobney's thread and sets its theme, which is information. Plant or human, information is culture, and that is what this project is about.

There came early on in Walnut Creek a firm decision: That if this project was to be truly about biodiversity and restoration, then all the work must rest on what is called local ecotype seeds. As we have seen, through the ages different species co-evolve or form patterns of relationships—symbiosis, niches, predator-prey dichotomies, adjustments to one another—that form the very notion of community. Species that have existed with one another for tens of thousands of years tend to do best when they continue to exist with one another. That is axiomatic. It is culture passed through generations of genetic information. Seeds pass on the legacy of learned relationships. The point, however, applies on an even finer level than species, to local ecotypes. Big bluestem exists throughout the tall-grass prairie, but big bluestem from Kansas differs substantially from big bluestem from Iowa. This is local ecotype, a fine-tuning of evolution to a specific location. Strains of a species develop certain immunities, resistances, and adaptations that allow them to survive in the specific conditions of place.

Biologists are only beginning to understand the sublime weight of this idea, but it is being applied across the board at Walnut Creek. As Drobney says, "Anything else would be just gardening." All of the seeds for the project, the packages of genetic information for several hundred species, are being gathered locally, specifically within a one-hundred-mile radius of the preserve. This is not restoration so much as it is resurrection.

Enter now the Iowa Prairie Network, both formal and informal. Enter now the culture of people. Had the federal government decided to pursue its usual course, it simply would have advertised for bids, signed a contract, and a load of seed would be trucked in, laid, and paid for. The truck and contractor would roll away and the bison and tourists would come. One, however, simply cannot buy a truckload of local ecotype seeds for Iowa tall-grass prairie. One must rely on a network of people to find them.

The basis of this existed as the Iowa Prairie Network, of which Drobney is a past president. It is a group of people around the state, normal people, not scientists, who for some reason or another have come to value tall grass and have become protective of, in many cases, a particular postage-stamp site. It is a growing strain of environmentalism throughout the West, a band of activists a former Montana congressman called "prairie fairies."

In the case of Walnut Creek, however, the demand for seed was so huge and the area it could come from was so small that the skeletal network was not good enough. The skeleton needed the flesh of the larger community, and Drobney needed to become more of a politician than a botanist.

To the corn farmers of Iowa, the tall grass is a symbol very much like wolves are a symbol to the cattle ranchers of Wyoming and Montana. Their ancestors filled poorhouses, insane asylums, and graveyards in a battle against the grass. They are not overly anxious to surrender the results of this war.

"I think it's going to be a big weed patch."

"Personally I think it's a disaster."

"They snuck in and bought up a whole bunch of this ground before we knew what was going on."

These comments came from neighboring farmers at public hearings setting up the project. Since then, the staff at the preserve has been trying to settle these enmities with such neighborly considerations as telling a farmer when his cows are out, closing gates, and reporting trespassers to landowners. It helps to know what matters in farm country, which is why most of the preserve's staff members are, like Drobney, native Iowans.

All of this, however, is a rear-guard action along the preserve's immediate borders. The real work spreads and goes much as in the following case.

Using satellite photos, Loren Lown, a former roofing contractor turned botanist, now the man in charge of Polk County Parks Department's postage-stamp prairies, found a relict piece of tall grass in a farmer's field. Lown visited the farm and confirmed the spot. He met the farmer, who was initially hostile out of fear of trespass. Lown took a walk with the man and identified some plants. He convinced him the site was unique, a legacy. He made friends and taught the farmer the names of the plants. Finally the farmer let him harvest seeds as long as it would not damage the site. Now the man knows what he has and wants it protected, and now Drobney has a few more bags of seeds.

On another site, Lown spent years working up to the farmer until he finally got the man to admit how much the hay was worth he annually cut from his patch of relict prairie.

"He said seven thousand dollars. Hell, I can pick seven thousand dollars' worth of seeds off that patch in two weeks," says Lown. So the feds leased the land from the farmer in exchange for seven thousand dollars' worth of prime alfalfa hay. This, too, is how one makes friends in farm country.

Pauline has a photo in the refuge offices, a clutch of temporary trailers directly downwind from a hog yard at the edge of the federal land. It is a 360-degree composite photo showing hundreds of people on almost the same site as the office. There are bunches of kids, farmers, and city people out from Des Moines for the refuge's first seeding, which occurred May 22, 1992.

"It was the birth of a prairie and these are the midwives," she says.

In the photo, some people are holding guitars, banjos, and a dulcimer. This is the band that the feds commissioned. A bluegrass band played and cloggers clogged and dancers danced, out on the bare soil. They spread the seeds then danced them into the ground.

Pauline and I and a couple of high school kids walked in that spot in the late fall of 1993, two growing seasons after the dancers' feet fell. It was to my eye a tall-grass prairie. Yes there were holes in the community. Many of the subtle relationships among grasses, forbs, fungi, insects, birds, and microorganisms will take centuries to be reestablish, but still this plot of a few hundred acres grew a dense mat of shoulder-high grass. There were at least forty species of plants.

I had questions for Pauline, but she was engaged by the high school kids, a couple of freshmen boys on leave from classes for the day to work their prairie project. I quizzed them a bit, and they seemed to know the plants. They had been gathering seeds and reestablishing their own private prairie plots near their homes. It was a project for school, for FFA, Future Farmers of America.

Farm kids? No, they said. They had been interested in wildlife, but in this part of Iowa the only way to pursue an interest in conservation was to sign up for FFA.

"Mr. Collins says FFA's not just for farm kids anymore," advised one of the students.

Pauline identified some foxtail barley growing round about. What about it, guys, how do you deal with foxtail? It's an aggressive, invading species of grass, which, like cheatgrass, prefers disturbed sites. The kids scratched their heads, probably running through a list of herbicides that might do in the offending invader.

Foxtail is the easy part, Pauline told them, because it likes disturbed sites and is an annual. You don't do anything about it. The natives will establish themselves and cover the bare ground, leaving no place for its seeds.

The kids liked the answer and went on identifying plants, no easy matter in November, when most of them are the same shade of brown, desiccated, leaves twisted, flowers gone. The kids were an integral part of the prairie network. Now there are prairie classes in virtually every high school and elementary school in eastern Iowa. When school kids go out rambling the old roadbeds and ditches, they spot relict patches of prairie. Someone else helps with the identification of the residents of these patches, and seeds are gathered. And genetic and cultural seeds are planted.

Pauline wants me to meet Gene Kromray, a businessman from Ottumwa who is president of the Iowa Prairie Network. It just happens that his truck pulls into the refuge's parking lot as we are talking about him, and he looks to be anything but a businessman, the operator of a computer consulting business. In white bib overalls, flannel shirt, and ball cap, he looks like the farmer he once was. Gene, who was at the refuge that day to deliver some seeds, is a dabbler, a godsend to the refuge. He developed the machine that harvests the tall-grass prairie's seeds, a leap in technology that will greatly speed the task ahead. The machine is, however, like most designs, not a clean leap at all but information drawn from culture. Gene gives me this story over lunch at a roadside café.

He's in his sixties and talks with his hands and a big part of his round farmer's face, still soft-spoken and easygoing. Gene has been instrumental in his county's roadside prairie program. In some midwestern states, about all that is left for public land is that portion of the road right-of-way not used as road, usually a strip on each side of thirty feet or so. Most of the time, it appears not even that much remains, because farmers, especially since machinery has become so large and difficult to turn, sometimes extend cultivation out across the right-of-way right to the edge of the blacktop. This was more or less okay with the counties, because it meant they didn't have to maintain the land, which usually meant seeding exotic grasses, mowing, and spraying herbicides on weeds.

Prairie advocates, however, have reclaimed some of that land and brought it back as native grasses, which require no mowing or spraying. They have used the roads as toeholds for habitat, often only after nasty legal action against those farmers who still insist on farming the right-of-way. More than once, a roadside prairie has been painstakingly reestablished, only to have the adjacent farmer blow on through with a field harrow and chemical sprayer.

Gene was a player in this battle and eventually became drawn to Pauline's cause, and all of this pointed to a crying need to find a better way to get seeds. The problem is, most seed harvesters—grain combines, for instance—cut and shatter seed heads, which in some cases might actually harvest prairie seeds, but would cause undue damage to those relict prairies that seeds must come from. Further, grain combines are large, heavy, and unwieldy—difficult to operate in the marginal terrain where relict prairies exist. If combines could operate there, there wouldn't be relict prairies.

Gene hit the library and found that the Australians had pursued an entirely different line of technology for harvesting seeds: devices, instead of shattering the head, stripped seeds, pulling them from the seed head the way a hand would. All the rest of the plant remained intact. With more research, he learned that this technology made it into the United States early in this century, particularly in Missouri, where it was used by a seed company, the Williams Brothers, to harvest its main product, seeds for Kentucky bluegrass. Bluegrass was all the rage then and farmers and ranchers were planting it farther west as fast as they could get the seeds. Literally hundreds of seed harvesters were bought by the farmers who raised and sold seeds. Eventually, the hub of the bluegrass business moved to South Dakota before it fizzled, and the harvesters went to ruin. Now and again, one shows up in a local parade, entered as a fairly interesting-looking piece of antique farm machinery.

Gene worked the phone and finally found a collector in Huron, South Dakota, who had one hundred of these harvesters. He drove the one-thousand-mile round trip and bought a couple of specimens. He hauled them back. His son, who is a machinist, went to work on necessary modifications. The tall-grass prairie is taller than bluegrass but still hosts a sort of understory, shorter plants that grow and produce seeds. Because of this, Gene and his son made their harvester by joining two bluegrass strippers, one to strip high and one low. The thing works, and he tows it all around Iowa, or at least within a one-hundred-mile radius of the refuge, participating in the resurrection. He works as a volunteer, mostly because he likes prairie, which was not always true.

"About ten years ago I got enough money together to buy an SLR camera and a macro lens. I started in the forest in April taking pictures and by the first of June I ran out of plants. I moved out into the ditches and I discovered that some of these plants there had names. I learned them, and I started visiting prairie preserves and now I've got twenty thousand images. I guess just being by myself out in these areas gets you hooked. I don't think that once you get it in your blood you ever get rid of it."

Pauline, the high school kids, and I are hiking across a soybean field rimmed by a ridge full of trees, mostly oak, walnut, and locust. The forest is too thick to walk through easily, so we choose the field, heading about a quarter mile across to an open, grassy break in the forest where a rare orchid still blooms in brown grass. I can't help but notice the trees are dying, one of the refuge's proudest achievements. It is not an easy task; one cannot simply cut these trees, because the stumps sucker and send shoots. So platoons of volunteers have fanned out across the refuge to girdle all the saplings, and they die. An army of happy volunteers killing trees: The refuge is taking the tree culture head on.

The record is clear on this. A survey was completed in 1847 that laid down the baseline conditions of this stretch of Iowa; there were no trees to speak of, certainly no forests, only single trees here and there. This far east this was not all pure prairie, but the edge of a subtle and, for humans, primal habitat type: the savannah, which is even less appreciated than grassland. There is deep irony here in that savannahs are indeed fundamental to the state of life we call human. Savannah, a rolling grassland sparred only here and there with a single scion of a tree, emerged during a period of global warming that thinned the forests of Africa. It was this shift that gave our little upright ape ancestors the lever they needed to prosper. Savannah is where humans prospered, the ecotome to which we are best adapted.

Savannah is only maintained by fire. The savannah trees set seeds, and these will grow to a forest unless fires sweep through every couple of years and kill the saplings. Before settlement, lightning and aboriginal people supplied the fire. The tree culture has suppressed those fires, believing the forest a good development. Settlers here chose sites near trees and let them spread.

I had spent the day before this walk with Loren Lown in an urban park he is restoring to savannah. It doesn't take long to learn to see the savannah hidden in the forest. In the mat of trees thick as an arm there are invariably a few trees centuries old, white, red, and burr oak, although the last is rare. Lown says the railroads preferred burr oak for ties, so those went west as the wood bed under the veins of steel.

That day, Loren was setting fires with a torch, trying to kill trees. "Burn, burn, you sons-o'-bitches," he'd say as he touched them off. He has been known to attack trees with a hatchet that shoots an herbicide under their skin.

As we set fires that day, the park's forester, who had been shanghaied to help control the fire, leaned on his rake, glared, and grumbled. Loren had fought a heavyweight political battle all the way to the county's commissioners for the right to burn and kill these trees, but the opposition had not yet surrendered. Opposition be damned. This death by fire is a part of this landscape's resurrection, and politics will not stop it. They burn.

We find the orchid, then work our way along the stream through the trees. Pauline stops at a spot and shows me the stream bed, about ten feet below the bank. This is not supposed to be. She has checked. Early settler accounts tell of splashing a horse and buggy through these streams, something that could not be accomplished down a ten-foot cut. All the streams have down-cut, formed a deeper channel, no doubt the result of grazing. Cows ate streamside vegetation, the roots that held the stream bed died, and the stream eroded its way down into the soil. Cows then kicked and trampled the banks, widening erosion's swath. On the uplands, farming practices allowed the black Iowa soil to wash downhill into stream-cut gullies. Some baseline research at the refuge showed erosion had deposited silt in draws as deep as thirty feet in some places, permanently changing the roll of the land.

To a botanist, this is no small matter. She shows me a meander in the stream where it dodges a high, south-facing bank. All of this configuration is no accident. South means drier, which in turn means different vegetation, different roots, and so a different course for the stream because the different roots change the path of resistance. What will happen now when the advancing forest has closed the canopy, shading out the dryland plants? Has the stream been altered? And what does one do now about the down-cut stream, which lowers the water table of the whole area? Can the same vegetation that made the stream what it once was now survive with the water six feet lower and the silt contour six feet higher?

How do we put all these genies back in their bottles when we are not altogether sure we have the right bottles? At Walnut Creek, these are not academic questions, but the job that lies ahead.

Loren's fire the previous day faced the same problem. Sure it was fire, but it fed on a denser, more mature forest than presettlement fires. This made it hotter. Might this not kill some seeds that would be otherwise stimulated by natural fire? How does one re-create nature in an unnatural world?

Loren showed me a clear patch of prairie, perhaps two acres in an area of forest. He's burning it now as grass, not forest, but a couple of years ago it was forest. He's let the sun in, and he was sure that was the right thing to do, but beyond that, he is still learning. Now the sun is here, and he simply watches the land year by year to see what happens, waiting for the next clues as to what the nature of this place might be. Because future fires will burn grass, not trees, they will be more like historic fires. From these future fires, we will try to divine the next step.

It's like a ratchet in that we take the first turn, then back off and let nature take a turn, as we listen and watch as closely as possible, knowing that we are not so much rebuilding nature in these places as nature is rebuilding us. Like genetic information, cultural information evolves. With each turn of this ratchet, nature reveals a bit more of itself to our science.

Pauline says her work is not simply an application of her knowledge as a botanist. It is not a prescribing of the known. Rather, it is exploration of the microscopically complex world we do not know and a realization that some of it is not knowable.

"This is a search for the unknown, and the unknown forms the model from which we work," she says. "We as a species are lessened by the loss of unknown places."

We drove all that day on straight gravel roads set in a grid across each other every mile and finally stopped to look at one of those. Pauline showed me where a bank had been cut and the roadbed built up, forever altering the soil profile below. The soil manifests itself as plants, as a botanist knows. The refuge will close these roads and bulldoze them back to something resembling the natural contour, but bulldozers cannot re-create the soil profile, and so for the ages, the plants above these buried roads will be different. The soil below will always contain different minerals in different layers, percolate water differently, and the plants will express all of this. Pauline believes the difference in plants may be visible even from aircraft. Someday one may fly above this refuge and still see the Jeffersonian grid sticking through the grass like the bones of a great buried beast.

In that the plants tell the history of a place, I'm not sure I would have it otherwise. One cannot cheat the land's library of its information.

In some real ways, the effort to re-find the landscape, which is the same as the effort to re-attach our lives to the land, is a religious quest, and it makes sense to consider it that way. Western literate culture, with its respect for authority, needed a creator and pursued one, eventually separate from the creation. This separation excused all manners of destruction.

Yet the creation remains, and some of us creatures need to find it. The creator and creation are the same and the unity is evolution. A place possesses a certain set of circumstances, weather, soil, and community, that lead to a certain manifestation of life in a place. Undeniably, the creation evolved certain parts of life to inhabit certain places, to show the nature of a place.

Certain cultural traditions took great pains to divine the intentions of the creation. They used shamans, who were not so much magicians or priests or herbalists or storytellers as they were seers, those who would travel beyond the veil to read the verities, to divine the creation's intentions. They saw and heard the mysteries that were just beyond every one else's ken.

I once heard a story of a man who had perfect recall, so he could never carry on a conversation. He had to live in isolation, because the merest stimulus, the merest sentence from outside his own head would recall everything. All the information in his head would come tumbling forth in a great rush, and he would be crushed by the pain of seeing.

I imagine that must be what it is like sometimes to be a botanist. I have been afield with many of them, and they are different, almost invariably quiet, distant. Undeniably, they see something different from what I see, as if the knowledge of the plants lifts a veil. The whole of it is there in the plants to be read, the full soul of a place, its life and the abuses of its life, the creation's intentions and the manifest violations of those intentions. Botanists are our shamans.

12

Agenda, Anti-Agenda

It is the nature of our culture to link ruminations on the past with prescriptions for the future, as if the future were to be dispensed from a safety-capped bottle bearing clear instructions. I mean first to begrudgingly observe this convention and then subvert it, a subversion particularly called for by the nature of the arid West.

A certain word seems an apt summary of the American present: adolescence. Stretching out behind us are a couple of hundred years of our collective childhood on this continent. I do not mean simply human; humans have existed here ten thousand years, the bulk of that time in relative peace with the landscape. The history of the continent before European conquest was not without upheaval, conflict, and stress, but it was by any reasonable standard, which is to say the standard of sustainability, a successful habitation. As it was constituted, it could have gone on forever; ours cannot.

By "our," I do not even mean European. That is a shorthand some have used to address some of the ugliness of the events of the place, but understanding demands more precision. Some Europeans, while not without their harmful effects, managed to come to the land, inhabit it, and still leave it able to survive. Some, particularly the French traders that founded lineages like the *metis,* melded into the existing culture, the authentic culture as it was formed by the land and the living manifestations of the land.

The real problem was industrial man. The brutality of Western industrialization re-formed the human race in waves belching from England first across Europe, then across this continent, across Japan, China, Russia, Eastern Europe, Central America, South America, India, and Africa. In all of these places it exacts a terrible toll in exchange for the undeniable material blessings it bestows, especially to a few increasingly able to insulate themselves from the toll.

Industrialism is rational and progressive; it has an agenda based on our assumption that we understand how things work. This very notion is dangerous to all landscapes, but it is particularly dangerous to the grassland West. The strictures of arid climate caused evolution to create a living face for the West that was superbly suited to survive conditions we did not and do not understand. The face reveals itself only slowly, over the centuries of cycles of drought, wind, and fire. It does not reveal itself to an impatient people with a Manifest Destiny, machines, and a guiding assumption that nature is just another machine.

The hubris of the industrial age was the belief that because we could make machines work, we could make the landscape into a machine and make it work like one. This thinking was naive and childlike and why I think of the word "adolescence." We are changing now to something some people call post-industrial. That's the way culture evolves. We can't go back; we have no choice in this. In this changing from our adolescence, we can either remain forever stuck in the naiveté of our childhood or we can attempt to gather some wisdom.

Probably we will do both, by degrees. Probably we will do one in some places and one in others, but there is no real reason to assume wisdom will prevail. To date, our society has shown no special inclination to mature. We still consider progress to be stoking the very economic engine that already has consumed so much of the planet's life as fuel. We are still able to rally the most "progressive" forces of our nation around candidates who say "It's the economy, stupid," when the fundamental issue is not the economy, it is life.

We are still adolescent enough to believe that our problems will be solved if Pop would only raise our allowance. We have been on a two-century binge, the sort of carousing fueled by the ability to deny mortality. Our society ran through the landscape like a hot rod full of teenagers full of beer.

When I was a teenager, my father told me several truths that I, like most adolescents, disregarded. Now they return in this context and seem relevant.

Only half flippantly, he said this: "Remember, I have taught you everything you know, but I haven't taught you everything I know." Only later I learned that all fathers have been saying this for all of time, but it's true. Now I take this thought as necessary to humble our science, to understand that everything we know was taught us by nature, but nature gave us brains evolved to their niche, so they are limited in their understanding. All that we know we have learned from nature, but we do not know all that nature knows.

My father said, "The world doesn't owe you a living." He meant this as a conservative axiom, my instruction in the work ethic. Today, all of our fine conservatives would agree with that statement, especially as it is applied to their image of welfare recipients and the poor in general, but if the world does not owe us a living, then why do we insist that every single piece of the world does? And why do we press the world to pay this imagined debt, far beyond its ability to pay? If the world does not owe us a living as individuals, why then does it owe us one as a species, no matter how many of our species we choose to add?

The last thing my father said went down the hardest, especially to a bright late-adolescent schooled in logic, in the power of law and a sense of justice. When my father would hand down an edict I thought particularly egregious, I would complain and demand justice, and he would always say, "I am your father, and I do not have to be fair."

Now we stand before nature demanding that it accommodate our sense of justice, that it treat each of us the same, that it not send us fires, floods, and storms and if it does, then it also send a government that insulates us from these. We demand that nature be kind to animals, and that we not die. Like Job, we demand justice from the creator, and the creator's answer was what my father paraphrased: "Where were you when I laid the earth's foundation?"

Yahweh, the God of grassland Bedouin herders, said this to Job and to us in describing the behemoth, the image of a terrible nature beyond our ken:

> *Look now at the behemoth which I made along with you;*
> *He eats grass like an ox.*
> *See now his strength is in his hips,*
> *And his power is in his stomach muscles.*
> *He moves his tail like a cedar,*
> *The sinews of his thighs are tightly knit.*
> *His bones are like beams of bronze,*
> *His ribs are like bars of iron.*
> *He is the first of the ways of God.*

An idea sets a standard for our future. I have seen it laid out best in an essay by David Orr called "Education and the Ecological Design Arts." In it he says the hallmark of the post-industrial age needs to be a new standard for machines, a post-industrial design process. Under such a system our artifice would become less crucial. Our machines, devices, and even relationships would be thought of not so much as whole-cloth inventions but as extensions or copies of nature. Orr wrote:

> Ecological design is the careful meshing of human purposes with the larger patterns and flows of the natural world, and the careful study of those patterns and flows to inform human purposes.
>
> Ecological design competence means incorporating intelligence about how nature works into the way we design, build and live. Design applies to the making of nearly everything that directly or indirectly requires energy and materials or governs their use including farms, houses, communities, neighborhoods, cities, transportation systems, technologies, economies and energy policies. When human artifacts and systems are well designed, they are in harmony with the ecological patterns in which they are imbedded. When poorly designed, they undermine those larger patterns creating pollution, high costs and social stress.

On its simplest level, this method of post-industrial design would amount to copying what nature has learned. Nature, through evolution, has carried out billions upon billions of trial-and-error experiments, and in the process must have hit upon some ideas we should know about. For instance, researchers recently found that some shellfish lay down the coating of their shells in a fashion that renders them super-hard, harder than any ceramics humans have been able to create, or at least could make until now. By aping nature's process, ceramic coatings have taken a quantum leap.

This process of copying nature is already heavily at work in the pharmaceutical industry, which sends researchers combing through the mind-boggling biological diversity of the tropical rain forests in search of useful natural formulations, instead of inventing medicine in the laboratory. This search places a new value on the accumulated wisdom of plants and of the people who study them. Currently, a quarter of all medicines in the United States come from material derived from plants, another 13 percent from microorganisms, and 3 percent more from animals. The biologist E. O. Wilson, however, believes this is only the beginning. He points out that only 3 percent of the flowering plants—5,000 of the 220,000 known species in the world—have been examined for useful compounds. What lies ahead? What of the grasslands? What of the several hundred species used by the people native there? How many of them have we already lost?

We have forever taken useful ideas from nature, but the concept of natural design expands to revolutionary scales. A couple of big ideas that conform to this standard already have been raised for the grassland West.

The bison are alive and well in pockets of resurrection, but there is an idea that would rocket the trend to a remaking of the face of the West. It comes, oddly, from geographers at Rutgers University, Frank and Deborah Popper, and usually is known as the Buffalo Commons. A similar but scaled-down idea has been floated that would apply to a vast stretch of eastern Montana's high plains, an idea called the Big Open. The Buffalo Commons is the same in principle but more ambitious in scale. It would apply to much of the open grassland states west of the 100th meridian, from Montana and North Dakota on south into Texas.

To the ranchers and boosters who have excoriated the Poppers, the idea is nothing more than one more eastern-based nut-case bureaucrat boondoggle fixing to grab their land, never mind that most of the land in question either is federal or is being paid for now with subsidies extracted from eastern-based bureaucrats. The idea is seen by its opponents as just another big national park, another playground for the rich. The Poppers, however, have something else in mind, a difference best explained by the fact that their idea is not so much a proposal as it is a description.

The Poppers are not saying that the people should be removed from millions of grassland acres to make a preserve for bison. They point out that depopulation is already occurring and will continue; depopulation is a fact. Their idea is to design a grassland economy based on its natural residents, particularly the bison.

As in the Sandhills of Nebraska, the land is reasserting itself throughout the region. The West that John Wesley Powell described always has been that West, despite our attempts to remake it. It took the old cowboy of the Turner ranch, Bud Griffith, long enough to realize that the way to herd buffalo was to figure out where they were going and go with them. It has taken us much longer to figure out the way to inhabit the West is just the same.

The Poppers do not prescribe a large park, but rather a commons, a vast and unfenced area where the bison and everything that went with them might roam. Simply, they see reestablishment of the great bison herds as the best use for a landscape that is being depopulated anyway. People, however, would be a part of this equation. There would be communities based on tourism, hunting, art, maybe even on plants as we begin to assess the commercial potential of native species. Mostly, however, it would be a bison economy based on meat and leather. It probably would not support a lot of people, and that is not a radical idea. Pre-settlement population of the plains was about one person per square mile, and despite the best efforts of the Homestead Act, that's on the order of the current population in the rural counties of the plains, away from the interstate highways and railroads. The farmers and ranchers are going broke and leaving; their land is being aggregated into ever larger holdings. These holdings could be further aggregated to cooperatively managed commons, a patchwork of open range across the West. It need not be a park or even federally owned. It can be cooperatively owned by private ranchers, by private organizations such as the Nature Conservancy, by eccentric media moguls like Ted Turner, by individuals, and by a combination of all of the above.

The idea, however, is not an agenda. It is a description. Because of this, grassland people speak about "when" instead of "if" the grass returns. In thirty years, the Ogallala aquifer will be gone. In less time, saline seep and salinization, erosion, and nitrate pollution will make huge areas of farmed land no longer farmable. Eventually, American taxpayers will cut the flow of subsidies to crop agriculture and the drought cycle will deepen, probably spurred by global warming. The grass will return. The question on the table is whether we can see this coming, accept it, and shape it.

Less developed, less inevitable, but potentially more sweeping and revolutionary—and certainly a firmer example of post-industrial design—is the work of Wes Jackson, the plant genius who has set for himself the business of reinventing agriculture. It is not enough to simply bring back the grass and the bison. We have an urban population, and it must be fed, and it must be fed grain. The culturing of grain is far and away the most damaging activity on the plains, but we ignore much of this damage because grain is, unlike most of our most destructive activities, bedrock necessary to civilization. Civilization is agriculture.

Wes Jackson, however, believes that all of agriculture has been built on one big mistake, one he intends to right. The central theme of agriculture is monoculture—that is, we raise early succession plants at the exclusion of all else. Early succession plants are evolved to move into a disturbed area and uniquely exploit the resources of the soil to pull from it maximum growth and energy in a single year. They are annuals, and the strategy of annuals is to put energy, literally its carbohydrates, into seeds. We and most other animals seek seeds, so our agriculture is simply a method to create disturbance. Put another way, agriculture is nothing but a determined effort to keep nature from taking its course. It is a locking of the system forever in immaturity, which is why it takes so much energy.

Jackson proposes reworking agriculture as a perennial polyculture. He proposes farm fields that copy the design of the mature communities of the prairie. First, because such a system is perennial, it would mostly eliminate tillage and along with it the energy that goes into the fuel tanks of tractors and, more important, the erosion that tillage engenders. Second, because the system is a polyculture, it would let the plants emerge as a community to do for one another those things we now do for them. They would, for instance, shield each other from predators, especially insects. A standing field of corn is an easy target for bugs that eat corn, simply because it sends out one big signal: corn. Once insects find it, no real trick in Iowa, they expend almost no energy moving from plant to plant. In a polyculture, the mix of signals makes the insects work for their living. Some plants even repel insects from their neighbors. A polyculture would stand a chance of achieving naturally the balance we now achieve with insecticides. As important would be the symbiotic cycling of nutrients a polyculture would allow, the very system that allows a prairie to, in Jackson's words, "sponsor its own fertility." The basis of the system would be a legume grown in tandem with a grass. Legumes fix free nitrogen from the atmosphere into the soil. The need for anhydrous ammonia ends, as does the nitrate pollution it engenders.

All of this thinking has gone well beyond thinking to the point that people now eat gama grass cakes in Salina, Kansas. That prairie town is the home of the Land Institute, founded by Wes and Dana Jackson in 1976 and engaged in research of his seminal idea ever since. In 1993, Jackson's work got a boost and some significant recognition in the form of a grant from the MacArthur Foundation, a so-called genius grant.

The work has proceeded along three lines: domesticating wild native perennials, domesticating an introduced exotic perennial, and breeding a perennial version of a plant now cultivated as an annual. The first approach led researchers, after examining three hundred species of perennials, to target three of those: eastern gama grass, Illinois bundleflower, and Maximilian's sunflower. The grains we eat now, corn, wheat, rye, barley, and rice, are domesticated wild grasses. Eastern gama grass is a part of that lineage and is thought to be an ancestor of the grass that gave us corn. It is, however, perennial. Illinois bundleflower is the legume in the mix because it fixes nitrogen. Maximilian sunflower is a prairie native, and native people ate its seeds.

The introduced perennial is a wild rye native to Bulgaria, Romania, and Turkey. The annual being bred to live as a perennial is grain sorghum. The Land Institute is breeding it with a close relative, Johnson grass, a native of Turkey now considered a weed in the southern plains, the stuff that spread along railroad beds during their construction.

Thus far, the research has established and successfully assembled the key pieces. That is, the combinations are feasible, they survive, and some produce seeds in quantities that are commercially viable and rival wheat fields. The research, however, has only begun. To succeed, the polycultures must mature as communities and maintain their yields indefinitely. The results so far are encouraging but not complete; the task is sufficiently ambitious to require generations of work.

What we must do in the West seems in some senses clear, at least when we take the big look that the place demands. There is a sort of manifest agenda and it reads like this: that we should first and always respect and maintain the grass. The West is grassland for a reason.

What that means first is coming to grips with the idea that the world does not owe us a living, at least some places. It is true that most of the grassland is so adapted to grazing that to cease grazing is to damage its health. Most but not all. By and large, the indicator of this, as it is for all else in the West, is moisture. The more rainfall a place gets, the more it can weather grazing, meaning the productivity of the land increases as one moves east from the Rockies. A second factor, however, impinges. The grass researcher William Lauenroth of Colorado State University says the second major indicator is a history of grazing. Long-term (by this he means at least since the Ice Age, maybe even twenty-five million years) grazing tends to select for species of grasses tolerant to grazing. These tend to be the shorter grasses, but especially those that reproduce with rhizomes and stolons, reproduce by shoots rather than by seeds. Grazed grasses are often deprived of their seeds.

Virtually all of the grassland east from the Rockies to the Mississippi has this long-term history of grazing. Virtually all of the grassland west of the Rockies—the Great Basin and the prairies of Washington and Oregon, the southwest deserts of Utah, Arizona, and New Mexico, and the Colorado River plateau—was not historically grazed by masses of large mammals. That region is heavily grazed today, and because of this, it is dying. Most of this grazing, especially in the Great Basin, occurs at the behest of the federal government, in that it occurs on public lands. The government, acting in our name, is wiping out an ecosystem as finely evolved and as vital as the Brazilian rain forests and the old growth of the Pacific Northwest. A reasonable agenda would ask that it stop.

This is not, economically, as radical a step as it sounds. For instance, Nevada, the cowboy state of Nevada that is grazed from one end to the other, holds only about one half of one percent of the nation's cattle. Their total disappearance would not so much as ripple the nation's economy.

A reasonable agenda, however, would not ask that all grazing stop. On lands where it is appropriate, it should proceed, but not as it has. Almost all of the public and private range lands in the West have been overgrazed or damaged by the peculiar habits of cattle and their European needs that evolved in places where it rains. This has been greatly magnified by the peculiar needs of Americans to dice property into small squares and hold it in fee simple. That is, we have fenced cows to increasingly smaller and smaller squares of landscape, concentrating their destruction. Historical grazing by bison improved the landscape because they had access to enormous areas. They grazed an area hard, then moved on, giving it months, more often years, to recover.

In all of this there emerges, then, a sort of ranking of preferences for change in grazing practices in the West, and the first preference is grazing by bison on huge chunks of open range, the buffalo commons. This is particularly true of the most arid reaches of the plains, the Great Plains and the short-grass steppe that is the last tier of grasslands before the rise of the Rockies: Montana, Wyoming, and Colorado. These are the lands driest and most susceptible to punishment, the lands once wholly given to bison.

A total reintroduction of bison is clearly impossible, at least in the short term, in that there are but 150,000 extant, not enough to populate the range and meet the needs of land and economy. The second preference of the land would seem to be grazing by cattle, but on large tracts of open range. Perhaps grazing by cattle could be made less harmful by further use of the old Spanish bloodlines, dryland cattle bred to the requirements of the West. Some of the longhorns remain, kept as rodeo stock and roadside attractions. The famous King Ranch of Texas experimented with hybrids and in the 1950s produced the Santa Gertrudis out of the old lines. These could more closely fit the needs of the range and have a place in the post-industrial design.

Above all, though, the land must avoid the plow, and this is where the assertions of the West come most squarely in conflict with our notion of progress. The conflict began in the nineteenth century when the spread of civilization's yeomen exterminated the nomad cultures of the plains. It continues today along the same battle lines when a new wave of civilization's progressives imposes an agenda. There is a solid band of vegetarianism mixed in the national environmental movement, and on its fringes, often even allied with the environmentalists, comes the animal rights movement. Both are firmly opposed to any expansion of an economy based on grazing, leather, and meat, as the economy of the grassland has been for ten thousand years.

In a naturalist's terms, the animal rights argument evaporates the quickest. The concept of right is a concept of law derived from civilization. Nature confers no rights, not even to humans. Rights are something we confer on one another as an expedient, the social contract. We, however, make our largest errors when we believe we are not a part but the whole of nature. Nature keeps order not with laws, justice, and rights, but with predator, prey, fire, and wind. Nature does not have to be fair. We were not there when nature laid the earth's foundations. Our every attempt to replace them with our own foundations has backfired, and nothing is better testimony to that than the face of the grassland West.

The killing of a deer is, in human terms, a sad and sobering event. I realize this because I have killed many of them. The planting of a grain field, however, is sadder. I know, because I've done that, too. I can understand that good people with the best of motives can believe that hunting and eating meat is wrong, but I will always think of the notion as city-bred. It is a product of distance. I used to think it was the product of a distance from hunting and the forest, but now I think it stems from a distance from and an ignorance of how we raise our grain. We are an agrarian culture and unable to admit that our wheat fields are among the most destructive forces on the planet.

Rights aside, though, there is a subsidiary and more formidable argument raised by vegetarians: the wisdom of eating low on the food chain. They assert—correctly and importantly—a vital message of post-industrial design: that it is necessary that our energy needs be met by current solar radiation. By this, they mean not just energy for cars, lighting, and home heating, but food. Eating grain and vegetables in general is simply a way of conserving existing solar energy.

A plant absorbs energy from the sun and stores it as carbohydrates. In some cases, especially with grain, we are given the choice of eating that plant or feeding it to livestock. In the former case, we use all of the energy the plant has stored. In the latter, however, the animal dissipates and uses about 90 percent of the energy in its own life forces. It stores but 10 percent, which is available for us to eat. Feeding grain to cattle wastes about 90 percent of the energy. True enough.

Modern agriculture, however, greatly complicates the argument, again by a factor of ten. Remember that the production, processing, and distribution of our food uses about ten calories of hydrocarbon energy to produce one calorie of food. In a sense, this only strengthens the vegetarians' argument. If we feed industrial agriculture's grain to beef, we use ten calories of stored plant carbohydrate and one hundred calories of hydrocarbons to produce one carbohydrate of meat energy. Exactly right as far as it goes, and a solid argument for never again pulling a package of burger out of the Safeway counter.

But what if I shoot a deer from grassland? If it's my own land

and I don't have to transport the deer, hydrocarbon energy drops to virtually zero. Grassland must exist. I can't eat grass. The deer ate it for me and in the process taught me to value the grassland with my life, and it will go on. My taking a deer from grassland means I take ten calories of energy for each calorie I consume. Under the American system of production, a person who buys a sack of potatoes at the supermarket takes ten calories of hydrocarbon energy for each calorie. Ultimately, what is more damaging to the life of the earth?

Now take the matter a step further. We feed no grain to bison, nor for that matter, do we have to feed grain to cows. Cows can survive perfectly well on grass alone and have for centuries produced meat by doing so. Grass-fed beef is, in fact, healthier meat, lower in fat and lower in the chemicals that now leach through our nation's grain supply. The habit of feeding grain to cattle developed with the growth of a nineteenth-century preference for rich food. It snowballed in this century with the stocks of surplus grain that began with the great plow-ups before and after World War I. It is the habit mostly responsible for Americans' suffering more heart disease than the rest of the world. We hid our grain surplus in cows to the point that 70 percent of our grain crop now feeds livestock.

What if we stopped? What if 70 percent of our cropland returned to grass and some of it raised cattle and some raised wildlife, or both? In his book *Food, Energy and the Future of Society*, David Pimentel estimates that ceasing to feed grain to livestock in the United States would free up as much as 130 million tons of grain annually, enough to feed four hundred million people. Ten acres of grass in Iowa could easily feed the same number of cattle that one hundred acres of grassland does now in Nevada. By retiring cropland to grassland, we have a place for the cattle and ranchers displaced from ceasing grazing in the Great Basin.

And what if the remaining cropland was less pressured to produce like a factory? The result easily could be less irrigation, fewer chemical fertilizers and pesticides, more organic agriculture, more no-till agriculture, and more room to experiment with Wes Jackson's permaculture.

In all of this there emerges an agenda of sorts, the outlines of a broad suggestion for the West. In a sense, it is a relatively simple matter. Believing the land equable and productive, we worked it too hard, whether the working was grazing or farming. The West has a simple definition of "too hard" based on rainfall. Historically, during deep drought cycles, the grassland has moved its plants east to follow the rain. The rain increases west to east and with it, the demands on the land also may increase.

Respecting the limits of the land would require a simple scaling back, a request that would not seem preposterous in light of nearly a century of grain surplus and waste. Pulling back would mean simply moving our most demanding activities east, cattle to wheat land, wheat land to corn land, grass to fill every hole left by the leaving.

This is only the skeleton of an agenda, but I will leave it at that for a number of reasons: First, the West is a diverse landscape and the details of an agenda will be specific to place, but more important, it is a lesson of the landscape and of its history that the very business of drawing an agenda is dangerous to the land.

Thomas Jefferson had an agenda for the land. So did Lincoln, and so did the Progressives, and they differed only in emphasis. It was the agenda of progress and democracy, and underneath all that the deeper agenda of settlement, prosperity, and stable communities, and underneath all that still, the agenda of security. It was a way of saying to the West that this corner of the world owes us a living, in particular a living that will ensure white-steepled churches and tree-lined streets. It was a way of saying that the logical and just God of Europe provided these blessings elsewhere, and if we are good people, will provide them here, under the terms of our contract with God.

The problem with this notion of the West's agenda is the failure to realize that we may not impose our ideas on the landscape, at least not without peril. An agenda is not imposed; it emerges. Nature does not decree; nature evolves. The deeper and real problem is that we failed to realize that the West is not ruled by a just and peaceful God.

I originally thought of this book as a manifesto, not without some acknowledgment of the irony involved. Manifesto is a term of certainty, even arrogance, and in the end, having considered the land, I am left more in doubt and ambivalence. Which is why I call this a manifesto.

The word "manifest" resonates in the history of the West, especially as Manifest Destiny. We meant by it that our charge was obvious, as if written on the land, when that was not the case at all. The only thing obvious about Manifest Destiny was our intention to impose our will, something quite different from the divine ordination the term implied.

Fittingly, though, "manifest" is a word at odds with itself. It means something that is evident and clear, that we need not force with artifice. Yet its Latin root means "struck by the hand," just the opposite. Manifest Destiny was true to this root's meaning in that the land indeed has been struck by our hand. So is the meaning of manifesto, by which we mean agenda.

Yet there are verities manifest on the face of the West, especially to those seers who through devotion and sensitivity have taught themselves to read them. It is from these and from this sense of the word "manifest" that a new manifesto must evolve.

Science has matured to understand that it must accommodate paradox and mystery, a realization vital to our understanding of the creation. We may begin an understanding of this as science did, with an idea as simple as a line. It helps here to understand what we all were told in high school geometry that a line is nothing but an idea, just as all of the formulae that drive our understanding of the world are abstractions. All of science forgot this, so we may be forgiven for doing so. To all of science and to pseudo-science —economics and sociology, for instance—the formulae, the abstractions, became the end, the reality. Our society is still Platonic in its idealism.

Only recently have some rebellious forces within the world of thought begun to consider a line in its reality, and it quickly became understood that real lines did not at all resemble the abstractions of lines we believed so necessary for our science. In some matters, this was not a critical distinction. We could predict the ultimate speed of a falling body with knowledge of formulae for gravitational pull and acceleration. Real-world infringements like atmospheric pressure were of minor consequence. In primitive matters like Newtonian physics, we understood the world well enough to get by.

But what of a line? The shortest distance between two points is a line of finite length, so the definition goes, but a real line has infinite length. The real distance between my house and the road has a couple of dips in it, so if I walk it, the length grows beyond the distance of line-of-sight to cover the dips. If an ant travels the same path, it encounters thousands of ant-scaled hills and valleys and the distance is greater still. It grows exponentially for a microorganism, a lifetime of travel of micro basin and range that I could walk in a minute and the ant could walk in a day. In reality all surfaces have these infinite variations, so there is no straight line from here to there. From this realization derived the study of fractal geometry and a blossoming began.

If a line could not describe an ant's path from here to there, then what good is the line, really? Now comes a second realization. The greatest share of our equations is what is known as linear, derived from the queen of sciences and rationalism. With the proper numbers cranked in, the equation, as revealed by the contemplation of the verities, worked its magic, and a predictable answer emerged. Usually, there was a pretty clear relationship between number in and number out. Slightly larger inputs yielded slightly larger or grossly larger outputs, larger still yielded again slightly larger or grossly larger still, depending on the equation. This was true for all manners of phenomena we considered: bodies at rest and in motion, squared legs and hypotenuse, yeomen and democracy, nitrogen and corn.

There existed, however, a separate class of equations called nonlinear, largely ignored because they were bizarre. They could not be graphed as continuous lines. A larger input might yield a larger input steadily for a while and then all of a sudden produce a smaller one. They were thought too nuts to consider until some rebels, from a number of disciplines, from physics to economics to biology, began studying the funhouse mirrors that were nonlinear equations. This gave rise in the 1970s and 1980s to what at first was called chaos theory and later bore less emphatic names. The emerging science got its push at first from the study of fluid dynamics, the swirls in liquids and gases such as eddies in rivers or the swirl of gasoline in a combustion chamber. It was the study of liquid movement in a world heretofore dry. It was the study of swirls, dips, and rises in a world heretofore given to lines.

It concluded that many vitally important forces of nature are essentially chaotic, not predictable in any real sense. A famous maxim of the field is what is known as the butterfly effect, a counter to the constant and so far unsuccessful attempt to predict weather. Beyond a few days, the weather in a given place is unpredictable, a stubborn fact that caused the linear thinkers to believe our data simply were not good enough. To the traditionalist, prediction was just a matter of getting better at the old science. The chaos people, however, suggested that just as the minute face of the terrain is relevant to the ant's perception of distance, so are small effects relevant to the weather. In a nonlinear world, minute inputs can have enormous consequences. To accurately predict the weather in Chicago next week, any formula would have to account for minute causes, on down to knowing the motion of a butterfly's wings in China.

At first this school of thought looked horribly nihilistic and dangerous to science, not to mention the careers of scientists, but that is only a first reading. A conclusion of chaos theory is that from it, nature does create order. A weird and elegant order emerges even in the simplest computer simulations of nonlinear equations. It is the same order that emerges in evolution from the random and chaotic levels of interaction with the building blocks of a cell. In this sense, then, chaos is the fundament of creativity. The agenda of nature cannot be decreed or even predicted; it will emerge in a pattern of its own devising.

The chaos theoreticians made some of their most interesting leaps, however, when they found that some of their results drawn strictly from abstract nonlinear equations, when graphed, superimposed nicely on graphs of such phenomena as the long-term price of cotton. Economics is not so much the study of markets as it is the study of human behavior. Now comes the assertion that this behavior is not nearly so straight-ahead and rational as we would like to assume, but as chaotic as the rest. Against this, how can there be a social agenda? Human behavior evolves.

The linear science and rationalism of the nineteenth century gave us the notion of progress. That's what linear equations are all about. It gave us a mechanistic world in which inputs of energy would make the machines turn, inputs of chemicals would make the plants grow, and inputs of free land would make democratic burgs spring and prosper all across the plains. We are beginning to understand that our productive systems, our factories, cannot run like machines. We have even taken some measure of wisdom from the plants to realize our farms are not machines. Still the social thinkers among us take some pride in calling themselves "progressives" and believe our social systems can be retooled like a machine. From this springs the requirement for a social agenda, which is nothing more than the drawing of a line.

Nature evolves from chaos, and the order is not ours to create, even in society.

Does this suggest we are helpless, or that we should abandon activism, attempts at social betterment, or creation of a human society to match the West? There arises from the chaos theoreticians an analogy, and it goes far in explaining what lies ahead for us in the West and the role of personal responsibility in achieving that future. John Wesley Powell, the first man to float the Colorado in a boat, would have understood the analogy instantly, and maybe he did. Maybe that's where he got the idea for a design of human occupation of the West that was based on nature. Maybe he would have understood it, because it is more than metaphor, derived as it is from the river and the fluid swirl of the eddies.

We, the humans of this place, have some say in our destiny the same way the pilot of a raft has some say in a raging whitewater river. Otherwise it would make no difference which pilot one chose to steer one down the Colorado. Anyone who has seen the parts of the Colorado that still rage would think such a notion insane. It does matter who is driving. Skill counts.

By the same token, however, no good rafter would say the raft goes where the rafter wishes. The river decides where the raft will go, in general. It's up to the rafter to decide a successful navigation of the river's agenda. To do this, one becomes an opportunist. One reads the river and pushes and pries when it will do the most good. Sometimes it is a matter of steering toward a river's more benign stretches. Sometimes the process is counterintuitive. By steering toward the scariest rocks, stacks, and holes, one takes advantage of the power right at their edge to push to a safe path. By this process we learn respect for ourselves and for the place that contains us. The river controls our destiny the same way nature does, but so do we. We feel not enslaved but freed by the river; otherwise, why would those people on those rafts laugh so much and obsessively return for more?

The grassland West is a place of freedom, a contention that can only be tested by parking beside a road one day, picking a spot on a horizon where there are no fences, and going for it, sometimes for days, sometimes inhaling its vastness, and sometimes, like the fractal ant, becoming lost in the minuteness of its detail. It expands infinitely without and within. It is a sort of freedom not contemplated by those who equated freedom, yeomen, and democracy, but a freedom unique to this place and one some of us now need.

We asked the West to make us a living, and it turns out this is not so much what we needed, or even what we asked. What we really meant was that it make some of us materially rich and secure, and the place is simply not geared for that, not without compromising its greatest gift, which is its ability to confer freedom. The point of the nineteenth-century agenda was to make us prosperous, but prosperity undermined the vitality of the West, and what kind of a trade is that? Some of us are not willing to make the trade, and we will arise as the true inhabitants of the West. We will accept reduced material circumstances in exchange for the privilege of the place.

The historian Donald Worster once wrote, in contemplating irrigation, that he would hate to see one more drop of water applied to his arid West, because one more drop of water would make it less free. Aridity is the soul of the place and water is the beginning of its enslavement. Its harshness is its beauty. Irrigation is compromising its freedom by arresting its motion, by making it safe and livable.

Freedom is a holy word of our culture and as such has lost its meaning. We no longer mean we want freedom; we want security, and they are very different goals.

The deconstructionist philosophers have an idea drawn from what two words once meant. The words are "fear" and "danger."

The word "fear" derives through old English from the German word for road and travel. In feudal times, in times of city-states, to travel, to be on the road, was to live in fear. Outside the city-state, one entered a no-man's-land, where the prince's protection did not extend. One lived in a city to gain that protection and to leave it was to live in fear.

Roads now mean something different to us, because travel has been industrialized. Roads now are simply extensions of the prince's power into the grassland. Travel, however, to leave the roads, to journey out of the protection of the prince, still implies a loss of security: to live in fear.

The word "danger" is derived from the French and meant to be under the protection of the lord. That is, it was the polar opposite of "fear." To be under the protection of the lord, which is to say under the protection of the city and civilization, was to understand that one had security only as long as the city conferred it. One lived literally at the sufferance of the lord, and a prince was free to lop off a subject's head as he chose. One lived in danger.

Given this, what choice do we have in the grassland but to journey and to live in fear? How else can one see its fine and crushing beauty?

13

Enclosure

On two successive days in Iowa and Missouri in November of 1993, I was given alternate visions of the grassland's future. I present both here because I believe they are not exclusive; both will obtain.

On the first day, I was bouncing along the blacktop grid roads surrounding Des Moines in a four-wheel-drive pickup with Loren Lown. He was showing me his projects, a network of postage-stamp and roadside prairies he had resurrected in his role with the Polk County Conservation Board. We drove to Polk City, an accomplishment not possible from some directions in the early days of settlement, even if four-wheel-drive pickups had existed. The area around Polk City had been a big wetland, impassable except when frozen in the winter; now it was drained by field tiles and farmed hard for corn and soybeans. The summer floods of 1993 had taken most of the corn, but it made no matter. The federal crop insurance payments would erase the memory of the floods, and the following spring, the planting would resume.

Outside Polk City there is a sand ridge, and the sand was the ridge's salvation. The ridge's flanks were too steep to plow. The ridge top, because it was sand, could be excavated all winter; sand holds no water and does not freeze. A place that can be dug all winter won't hold water for corn, but it will hold coffins, and so becomes the cemetery of a pioneer town. The ridge erects a hill, not much of one, but then it doesn't take much of a hill to stand out in Iowa. The tallest point between that graveyard and Minneapolis 250 miles to the north is the grain elevator at Bradford, Iowa. By the 1840s, the Indians had been chased from hereabouts and the cemetery began filling with pioneer graves.

The earliest generations of headstones mark single graves of children, with no family around. These were the kids of westbound settlers who had paused here only long enough to leave a child to the land's safekeeping. Also among the earlier stones, there is a family plot, one stone for a man whose life was used up by the land; next to him, three wives even harder used; and stretched out beside him in succession and around them eighteen children, like cordwood. Back farther still, in a rougher corner of the cemetery, there are rows of simple rectangular stones, uninscribed. The graveyard took the poor, too, but there was no money for carving the identity of the people whose lives on surrounding farms had not left them enough to ransom a name from time.

Today, the unburied of Polk City pay a man to keep the cemetery trimmed, and he does, by maintaining a clipped carpet of Kentucky bluegrass around the stones, a sort of monument in its own right.

Around the edges of the hilltop, however, on that forgotten ground too steep for corn or caskets, a prairie is rising. No one cared about the land on the flanks of that ridge for more than a century, and without fire, it went back to trees. The forest buried the tall-grass prairie at its feet. Then a few years ago, a woman of Polk City, a daughter of the stones, found the stunted, etiolated remnants of Indian grass, leadplant, hairy grama, side oats, and dropseed surviving in the trees. She organized and lobbied. Soon there was a fire, and the prairie returned.

A day's drive northwest of Polk City at Wounded Knee, South Dakota, there is another graveyard that the Iowa cemetery brought to mind. There, "rest in peace" seems a cynical joke. Among crosses and medals there are medicine bundles, tobacco offerings, and smudge, the markings of another prairie hilltop capped with the leavings of a brutal century. This hill holds the victims of the massacre at Wounded Knee in December of 1890 when the Hotchkiss guns of the U.S. cavalry massacred at least 153 and as many as 350 men, women, and children, Big Foot's band of Sioux.

Wounded Knee is full of angry graves bit by years of harsh Dakota wind, prickly pear, grit, and gravel. Like the rest of the West, the land around Wounded Knee is used hard. If these graves are to ever settle their anger, it will settle only when the prairie rises again.

The graveyard prairie of Polk City is sufficiently recovered now to raise seeds that are spread through the state's prairie network. Some go to the eight-thousand-acre restoration project at Walnut Creek, some twenty miles away. The graveyard prairie raises seeds of big bluestem, little bluestem, compass plant, leadplant, prairie clover, ground cherries, Indian grass, and in these are seeds of dickcissel, Henslow's sparrow, horned lark, fox snake, elk, wolf, and Sioux.

The old farmers around here, the hard-heads, call the prairie "weeds." A botanist I know tells an old farmer that a weed is just a plant out of place.

"Are you telling me these plants have a place?"

Just so. The prairie raises a place.

The plains tribes, each in their own language, usually called themselves by a word that means "real people." We the educated have generally interpreted this as a sort of primitive arrogance, but hear this. As some of the tribes were forced from their home lands and onto reservations many times hundreds of miles and whole biomes away, many of the tribes ceased using their names. That is, once out of place they no longer were "real people."

The seeds of the Polk City graveyard resurrect the prairie and the prairie resurrected the Lakota, in the sense that we all ought to be Lakota, which is to say, we ought to be able to call ourselves, in our own language, real people. The seeds raise the possibility of a people no longer out of place, the possibility that we may again have a name, that we may ransom that name from the future.

Only late that afternoon after the graveyard did it occur to me that Kansas City was so close, maybe three hours by interstate southwest, and only late in that day did it occur to me that this was where I needed to go to finish my journey. I had to catch a flight the next afternoon, so I arose the next morning at 4:30 and beat a rental car down the deserted highway into Missouri. At Kansas City in the pouring rain and rush hour, I crossed the Missouri, the river whose other end had started me wandering through the grassland. Here, though, it looked nothing like the mountain streams I know, and I crossed it without much regard. It was not my destination. I was headed for Independence, a few miles east down another interstate. I already had passed through Liberty. Jefferson City lay a few miles east still, but I was headed for Independence and had a map.

I left the interstate there and was flung straight into a sprawl of suburban Quick Stops, fast food, and mini-malls that make the face of our nation that most people know. Then I drove a couple of miles toward the edge of the grid to where the newer rambling, wood-sided homes are chewing up the countryside with bluegrass lawns and cookie cutter architecture, and continued on around a winding road to Jackson County's Fleming Park. It was still early, and the park was empty of people, and besides it was raining a lot harder than a Montanan can imagine rain.

I found my way around Lake Jocomo, which is really a reservoir (contrary to the practice of the federal officials who build reservoirs and letter maps, we ought to call reservoirs by their names). The road topped a hill in the woods and some deer jumped across the road, then the woods petered out and surrendered the roadside to a long stretch of woven wire fence that contained a close-clipped exotic grass pasture. Inside stood standard issue, roadside-attraction bison, outside an elevated platform to provide the vantage to camcorders.

I had not driven all this way, though, to see these bison, so I drove on to the next fenced pasture of what a sign had just informed me was the "Native Hoofed Animal Enclosure." It really was a zoo, apparently another name too harsh for the park's sign. An adjacent sign said: "Please feed only apples, pears and corn. No other foods are allowed."

Inside were the elk I had come to see. Another sign said one of them, a bull, was from Montana and had been granted its life here by the governor of my home state, as if an elk's life was something a governor could grant. Briefly the sign told the story of the Sweet Grass Hills and an elk that wandered down the Missouri to be captured here, 1,800 miles from home.

Behind the fence in another close-cropped pasture fanned a herd of cows and calves, and, off alone serenely ruminating next to a pond, was bedded an enormous bull. He was the only bull in sight and had to be the elk I wanted to see, but I couldn't be sure and had to know. I drove back across the park to a restored colonial farmhouse that was park headquarters and asked.

"No, that's not him. That's the old bull, and he's just livestock by now. He's been raised all his life here and tame as came be. It's the rut, and we have to separate the Montana bull and a few cows from the old bull with his cows. Otherwise that Montana bull would try to kill the big guy. He'd kill him."

I'd always thought that the urge that brought that bull this far was the urge to spread his genetic information, and I believed he would indeed kill the big bull for the privilege of doing so.

I asked the guy if he didn't think it a shame to have a set of genes that can carry an animal 1,800 miles locked up behind a fence, and he said it was, but what else were they to do. Leave him wandering round the suburbs, and he'd be hit and killed on a road sooner or later.

Now I drive back to the Hoofed Animal Enclosure, determined to see the bull. I walk the perimeter of the front pasture back to where I can see the cross fence separating the two groups of elk. Every span of wire between every post in the cross-fence, once strung straight, has been beaten into an arc by the elk from Montana. The guy at park headquarters had told me the Montana bull spends the whole rut either breeding or attempting to batter down that separating fence so he can kill the other bull and take his cows. The other bull had not responded to the upstart; no challenges issued from the opposite direction. Every U-shaped arc in every span of wire is pointed toward the old bull.

The back pasture is not cleared like the front, but choked with sapling trees, so I slowly crawl up and down the perimeter fence in the rain peering into the trees with my binoculars. Now and again I catch sight of a cow or maybe a set of legs that could be a bull, no telling.

Often in fall in Montana I will walk for whole weeks in the mountains, in late September when most everyone else is gone from the back country. It is the best time of year to be there for a panorama of reasons, but mostly I go to hear the bugle of bull elk. Like the cry of migrating geese, it is one of life's most sublime sounds, primal and haunting. One expects these barrel-chested, hormone-crazed beasts to issue something in the bass range, but the bugle of an elk, a song they sing almost only in the rut, comes from another place. It is a flute's whistle, shriller than anything one would expect, and it is without analog.

Only the most persistent, obstreperous, and biggest of the bull elk breed. After a month or so of fighting and copulation, a bull is worn out, almost literally. In a tough winter the big bulls die before any other animals in the herd, so weakened are they by their exertions. Maybe they know this, knowledge that tunes the shriek of their bugles.

The rutting cry of a bull elk is a seed song, a grass song, strung on the sad, tense line between mortality and immortality. I have always heard it as such, but never so sad as on that rainy morning when it broke through the sodden air of suburban Independence, Missouri, and filtered to my ears through a straight stretch of woven-wire fence.

Acknowledgments

Books are not the sole property of their authors, but flow from the work of others, not the least of which are other authors. In recent years, the plains have been enriched by a retelling of their many stories, and some of those steered me toward this work. Among those are Ian Frazier's *Great Plains*, William Least Heat Moon's *PrairyErth*, Evan S. Connell's *Son of the Morning Star*, and Kathleen Norris's *Dakota.* Although not specifically about the plains, Gary Nabhan's *The Desert Smells Like Rain* also provided a key philosophical nudge by suggesting that we as a people are a manifestation of the plants that surround us.

Beyond these, specific information and guidance for the book came especially from the works cited in the bibliography and from the many people who sat for my questions during the months of this writing. The latter I try to identify and acknowledge in the text in that this is their story, too.

It also is the story of several unnamed informants and guides. I drew much information and assistance from one of Montana's best botanists and teachers, Keith Shaw, and from his wife, Leslie. Their love of and devotion to the natural world was as much an underpinning of this book as any of the above-named works. As helpful with information and with opening their many fine facilities to me were the staff of the Nature Conservancy. The organization's work on restoration of the plains is invaluable.

Beyond the obvious personal debt, I owe a professional one to my mother-in-law, Audrey Stone. Her voracious reading habits serve me as a de facto clip service.

Andy Kulla, a botanist in the Lolo National Forest, steered me toward several key pieces of information, but more importantly, toward an appreciation of native plants in the first place. So did the botanist John Pierce, who helped me find the local ecotype root stock that became the basis for my personal restoration project of bunchgrasses on my own land. Much of this book flowed from that experience.

I am especially indebted to my friend the historian Dan Gallacher, who reviewed the manuscript and provided his helpful comments. The manuscript also drew great help from its editor, Caroline White, and I thank her for her good work.

Finally, and most importantly, the writing life is not easy, probably most difficult for those people sentenced to live closest to the writer. Yet this particular writer's life is a good one, largely because of the strength and support of the person closest to me, my wife, Tracy.

Bibliography

Adams, William. "Natural Virtue: Symbol and the Imagination in the American Farm Crisis." *Georgia Review* 39 (4 1985): 695–712.

Agenbroad, Larry D., Jim I. Mead, Lisa W. Nelson, eds. *Megafauna and Man: Discovery of America's Heartland*, vol. 1. Hot Springs, S.D.: Mammoth Site of Hot Springs, 1990.

Archer, Sellers, and Clarence Bunch. *The American Grass Book*. Norman, Okla.: University of Oklahoma Press, 1953.

Brown, C. Allan. "Thomas Jefferson's Poplar Forest: The Mathematics of an Ideal Villa." *Architectural Digest* 10 (2 1990): 117–39.

Brown, Dee. *Bury My Heart at Wounded Knee*. New York: Holt, Rinehart & Winston, 1970.

Brown, Lauren. *Grasslands*. New York: Alfred A. Knopf, 1985.

Brown, Lester R. *State of the World 1993*. New York: W. W. Norton, 1993.

Browne, Malcolm W. " 'Dwarf' Mammoths Outlasted Others." *The New York Times*, March 25, 1993.

Bruggen, Theodore Van. *Wildflowers, Grasses & Other Plants of the Northern Plains and Black Hills*. Rapid City, S.D.: Badlands Natural History Association, 1992.

Cather, Willa. *My Ántonia* (1949 ed.). Boston: Houghton Mifflin, 1918.

Chaney, Ed, Wayne Elmore, William S. Platts. *Livestock Grazing on Western Riparian Areas.* U.S. Environmental Protection Agency, 1990.

Chatwin, Bruce. *What Am I Doing Here?* New York: Viking, 1989.

Connell, Evan S. *Son of the Morning Star.* San Francisco: North Point Press, 1984.

Cunningham, Isabel Shipley. *Frank N. Meyer: Plant Hunter in Asia.* Ames, Iowa: Iowa State University Press, 1984.

DePalma, Anthony. "The Mexicans Fear for Corn, in Danger from Free Trade." *The New York Times,* July 12, 1993, A-1.

Devine, Robert. "The Cheatgrass Problem." *The Atlantic Monthly,* May 1993, pp. 40–45.

Dormaar, J. F. "Waiting for a Vision." *The Explorer's Journal* 66 (4 1988): 150.

Egan, Timothy. "Wingtip 'Cowboys' in Last Stand to Hold on to Low Grazing Fees." *The New York Times,* October 29, 1993, A-1.

Ewing, Sherm. *The Range.* Missoula: Mountain Press, 1990.

Fagan, Brian M. *The Great Journey: The Peopling of Ancient America.* London: Thames and Hudson, 1987.

Frazier, Ian. *Great Plains.* New York: Farrar, Straus & Giroux, 1989.

Gale, Donn A. Reimund, and Fred Gale. *Structural Change in the U.S. Farm Sector, 1974–87.* U.S. Department of Agriculture, 1992, Bulletin 647.

Gilmore, Melvin R. *Uses of Plants by the Indians of the Missouri River Region,* Bison Book ed. Lincoln, Nebr.: University of Nebraska Press, 1977.

Gleick, James. *Chaos.* New York: Viking Penguin, 1987.

Goetzmann, William H. *Exploration and Empire: The Explorer and the Scientist.* New York: Alfred A. Knopf, 1966.

Gould, Stephen Jay. *Bully for Brontosaurus.* New York: W. W. Norton, 1991.

Gribbin, John and Mary. *Children of the Ice.* London: Basil Blackwell, 1990.

Haynes, C. Vance, Jr. "Curry Draw, Cochise County, Arizona: A Late Quaternary Stratigraphic Record of Pleistocene Extinction and Paleo-Indian Activities." *Geological Society of America Centennial Field Guide,* Cordilleran Section (1987).

———. "Geoarchaeological and Paleohydrological Evidence for a Clovis-

Age Drought in North America and Its Bearing on Extinction." *Quaternary Research* 35 (1991): 438–50.

Heat Moon, William Least. *PrairyErth*. Boston: Houghton Mifflin, 1991.

Higgins, Kenneth F. "Interpretation and Compendium of Historical Fire Accounts in the Northern Great Plains." Washington, D.C.: U.S. Department of the Interior, Fish and Wildlife Service (1986).

Hudson, Elizabeth. "Again, a Home Where the Buffalo Roam." *The Washington Post*, September 19, 1993, A-1.

Hyams, Edward. *Plants in the Service of Man: 10,000 Years of Domestication*. New York: J. B. Lippincott, 1971.

Jackson, Donald. *Thomas Jefferson & the Stony Mountains: Exploring the West from Monticello*. Chicago: University of Illinois Press, 1981.

Jackson, Wes. *Altars of Unhewn Stone*. San Francisco: North Point Press, 1987.

Kenworthy, Tom. "Battle over Grazing Fee Plan Pits Visions of 'Old' vs. 'New' West." *The Washington Post*, October 31, 1993, A-6.

———. "Big Livestock Operators Dominate Public Range." *The Washington Post*, April 29, 1993.

———. "Information Scarce on Range Lands." *The Washington Post*, January 8, 1994.

Knobel, Edward. *Field Guide to the Grasses, Sedges and Rushes of the United States*. New York: Dover, 1977.

Lattimore, Owen. *The Mongols of Manchuria*. New York: Howard Fertig (John Day reprint), 1969.

———. *Mongol Journeys*. New York: AMS Press, 1975.

Lowdermilk, W. C. *Conquest of the Land Through 7,000 Years*. U.S. Department of Agriculture, 1953, Bulletin 99.

Luoma, Jon R. "Back Home on the Range?" *Audubon*, March–April, 1993, p. 54.

Madson, John. *Where the Sky Began: Land of the Tallgrass Prairie*. San Francisco: Sierra Club Books, 1982.

———. "On the Osage." *Nature Conservancy Magazine*, May–June 1990, pp. 7–15.

Martin, P. S. "Vanishings and Future of the Prairie." *Geoscience and Man* 10 (1975): 39–49.

Martin, Paul S., and Richard G. Klein, eds. *Quarternary Extinctions: A Prehistoric Revolution*. Tucson: University of Arizona Press, 1984.

Mathews, Anne. *Where the Buffalo Roam*. New York: Grove Weidenfeld, 1992.

McHugh, Tom. *The Time of the Buffalo*. New York: Alfred A. Knopf, 1972.

Milchunas, D. G., and W. K. Lauenroth. "Short Grass Steppe." In *Ecosystems of the World*, ed. Robert T. Coupland. Amsterdam: Elsevier, 1991, pp. 183–226.

Milchunas, D. G., O. E. Sala, and W. K. Lauenroth. "A Generalized Model of the Effects of Grazing by Large Herbivores on Grassland Community Structure." *The American Naturalist* 132 (1 1988): 87–106.

Milstein, Michael. "Politics May Destroy Parks." *High Country News*, June 15, 1992, pp. 12–13.

Norris, Kathleen. *Dakota: A Spiritual Geography*. New York: Ticknor & Fields, 1993.

Pielou, E. C. *After the Ice Age*. Chicago: University of Chicago Press, 1991.

Polunin, Nicholas. *Introduction to Plant Geography*. London: Longmans, Green, 1960.

Quayle, William A. *The Prairie and the Sea*. New York: Eaton and Mains, 1905.

Reisner, Marc. *Cadillac Desert*. New York: Viking Penguin, 1986.

Rice, Peter M. "Sulfur Cinquefoil: A New Threat to Biological Diversity." *Western Wildlands*, Summer 1991, pp. 34–40.

Rifkin, Jeremy. *Beyond Beef: The Rise and Fall of the Cattle Culture*. New York: Dutton, 1992.

Rölvaag, O. E. *Giants in the Earth*, transl. Lincoln Colcord. New York: A. L. Burt, 1927.

Safire, William. *The First Dissident: The Book of Job in Today's Politics*. New York: Random House, 1992.

Sandoz, Mari. *The Buffalo Hunters*. Lincoln, Nebr.: University of Nebraska Press, 1954.

———. *The Cattlemen*. Lincoln, Nebr.: University of Nebraska Press, 1958.

———. *Love Song to the Plains*. Lincoln, Nebr.: University of Nebraska Press, 1961.

———. *Old Jules*. Lincoln, Nebr.: University of Nebraska Press, 1935.

Soule, Judith, and Jon Piper. *Farming in Nature's Image: An Ecological Approach to Agriculture*. Washington, D.C.: Island Press, 1992.

Soule, Michael E. "The Onslaught of Alien Species, and Other Challenges in the Coming Decades." *Conservation Biology* 4 (3 1990): 233–39.

Stauffer, Helen Winter. *Mari Sandoz: Story Catcher of the Plains*. Lincoln, Nebr.: University of Nebraska Press, 1982.

Stefferud, Alfred, ed. *Grass: Yearbook of Agriculture*. Washington, D.C.: U.S. Department of Agriculture, 1948.

Stegner, Wallace. *Beyond the Hundredth Meridian* (Penguin 1992 ed.). New York: Houghton Mifflin, 1954.

———. *Where the Bluebird Sings to the Lemonade Springs*. New York: Viking Penguin, 1992.

———. *Wolf Willow*. New York: Viking, 1955.

Stevens, William K. "The Heavy Hand of European Settlement." *The New York Times*, August 10, 1993, B5–8.

———. "Home on the Range (Or What's Left of It)." *The New York Times*, Oct. 19, 1993, B5–8.

———. "Restoring an Ancient Landscape: An Innovative Plan for the Midwest." *The New York Times*, March 2, 1993, B5–6.

———. "A Snapshot of the Continent's Flora, Invaders and All." *The New York Times*, October 5, 1993, B5–8.

———. "Want a Room with a View? Idea May Be in the Genes." *The New York Times*, November 30, 1993, B5–9.

Stubbendieck, James, Stephen L. Hatch, and Charles Butterfield. *North American Range Plants*. Lincoln, Nebr.: University of Nebraska Press, 1981.

Temple, Stanley A. "The Nasty Necessity: Eradicating Exotics." *Conservation Biology* 4 (2 1990): 113–17.

Toole, K. Ross. *Twentieth Century Montana: A State of Extremes*. Norman, Okla.: University of Oklahoma Press, 1972.

Trimble, Donald E. *The Geologic Story of the Great Plains*. Washington, D.C.: U.S. Department of the Interior, 1980.

Tyser, Robin W., and Christopher A. Worley. "Alien Flora in Grasslands Adjacent to Road and Trail Corridors in Glacier National Park." *Conservation Biology* 6 (2 1992): 253–61.

Walker, Mildred. *Winter Wheat*. New York: Harcourt, Brace, 1944.

Warrick, Robert. *Wheels of Fortune*. Lincoln, Nebr.: Center for Rural Affairs, 1976.

Weaver, J. E., and F. W. Albertson. *Grasslands of the Great Plains*. Lincoln, Nebr.: Johnsen, 1956.

Weymouth, Lally, ed. *Thomas Jefferson: The Man, His World, His Influence*. New York: G. P. Putnam's Sons, 1973.

"Where Breakdown and Bankruptcy Play." *The Economist*, November 2, 1991, pp. 21–23.

Wilford, John Noble. "Clues to Earliest Americans in 11,700-Year-Old Campsite." *The New York Times*, March 25, 1993.

———. "Remaking the Wheel: The Evolution of the Chariot." *The New York Times*, February 22, 1994, B5–7.

Wilson, Edward O. *The Diversity of Life*. New York: W. W. Norton, 1992.

Worster, Donald. *Rivers of Empire*. New York: Oxford University Press, 1985.

———. *Under Western Skies: Nature and History in the American West*. New York: Oxford University Press, 1992.

———. *The Wealth of Nature: Environmental History and the Ecological Imagination*. New York: Oxford University Press, 1993.

Wuerthner, George. "Some Ecological Costs of Livestock." *Wild Earth*, Spring 1992, pp. 10–14.

Zavalete, Erika. "Ruminating on Range Reform." *Inner Voice* 5 (4 1993): 1–7.

Index